Jar Tests for Water Treatment Optimisation: How to Perform Jar Tests – a handbook

Jar Tests for Water Treatment Optimisation: How to Perform Jar Tests – a handbook

Martin Pivokonský, Kateřina Novotná, Lenka Čermáková and Radim Petříček

Published by

IWA Publishing
Unit 104–105, Export Building
1 Clove Crescent
London E14 2BA, UK
Telephone: +44 (0)20 7654 5500
Fax: +44 (0)20 7654 5555
Email: publications@iwap.co.uk
Web: www.iwapublishing.com

First published 2022
© 2022 IWA Publishing

Disclaimer

British Library Cataloguing in Publication Data
A CIP catalogue record for this book is available from the British Library

ISBN: 9781789062687 (Paperback)
ISBN: 9781789062694 (eBook)
ISBN: 9781789062700 (ePUB)

This eBook was made Open Access in March 2022

Contents

About the authors

Martin Pivokonský is a distinguished Czech researcher and an expert in the field of water treatment. He graduated in environmental engineering from the Institute for Environmental Studies, Faculty of Science, Charles University in Prague, where he also obtained a PhD and later became an associate professor. Currently, he is a director of the Institute of Hydrodynamics of the Czech Academy of Sciences (Prague, Czech Republic), and he also leads a research team focused on drinking water treatment at the Institute. He has been a supervisor for many MS and PhD students. He has vast experience in the optimisation and design of treatment technologies for practice, and provides expertise and consultancy in the field of drinking water treatment.

The co-authors **Kateřina Novotná**, **Lenka Čermáková**, and **Radim Petříček** are all postdoctoral researchers at the Institute of Hydrodynamics of the Czech Academy of Sciences, where they are members of the research team focused on drinking water treatment.

Kateřina Novotná obtained MS degree from the Institute for Environmental Studies, Faculty of Science, Charles University in Prague, and PhD degree from the Department of Water Technology and Environmental Engineering, Faculty of Environmental Technology, University of Chemistry and Technology Prague. Her professional focus involves characterisation and coagulation-based removal of natural organic matter during drinking water treatment, and the occurrence of anthropogenic pollutants in water.

Lenka Čermáková conducted her MS and PhD studies at the Institute for Environmental Studies, Faculty of Science, Charles University in Prague. Her research interests within the field of drinking water treatment include coagulation–flocculation of natural organic matter, and also adsorption of both organic and inorganic impurities onto activated carbon and innovative adsorbents.

Radim Petříček obtained MS and PhD degree from the Department of Chemical Engineering, Faculty of Chemical Engineering, University of Chemistry and Technology Prague. His research focus is on the removal of various impurities during drinking water treatment by coagulation–flocculation with particular interest in aggregate formation and properties. He also deals with construction and utilisation of pilot-plant devices.

doi: 10.2166/9781789062694_ix

Preface

This publication is designed to be used as a handbook providing instructions for the proper method of conducting jar tests, which are needed for the investigation and optimisation of the coagulation–flocculation process, a key step of drinking water treatment. Coagulation–flocculation has great potential to remove a wide range of undesirable impurities if operated accurately. However, optimisation of this process might appear difficult since it depends on many factors, including the nature of the target impurities and the overall composition of raw water. To determine the most suitable operational parameters of coagulation–flocculation that result in the best possible treated water quality, jar tests are a necessary nuisance. Here, an essential theoretical background is provided together with detailed practical instructions for properly conducting jar tests. Unfortunately, the effects of some critical coagulation–flocculation parameters are often overlooked in drinking water treatment practice and even in some laboratory studies that investigate the process via jar tests, and we believe that this book will contribute to solving this problem. The handbook is therefore intended for everyone who is interested in conducting the jar tests in a way that will provide the best possible results that will be comparable among practitioners. The audience may therefore include employees of drinking water treatment plants as well as students and starting researchers.

Foreword

I have had the pleasure of collaborating with Martin Pivokonský and his team for over 5 years, largely on the topic of coagulation and flocculation, which forms the focus of this book. Coagulation and flocculation is a critical water treatment process employed to remove a range of particles, colloids and dissolved material. On the face of it, this process can appear deceptively simple; however, optimising operating conditions for what can be a highly variable range of water quality conditions is actually a highly complex process. This book draws from both an extensive depth of knowledge developed by the authors through their numerous research endeavours that have explored and identified the underlying mechanisms driving floc formation, as well as a thorough analysis of both contemporary and long established literature, enabling them to make the latest understanding of the theory of coagulation and flocculation accessible. The key aspects include the use of novel coagulants and a detailed analysis of the impact of coagulation conditions and water quality on floc properties, and in turn the impact on downstream separation processes. However, this book moves beyond the theoretical to describe the more practical aspects of how to implement the jar tests that are used to simulate coagulation–flocculation within the laboratory, recommending approaches to ensure that common mistakes are avoided. For example, the laboratory equipment required to ensure a reflection of the full-scale process as accurate as possible is described; simple but detailed step-by-step instructions are provided that will enable the novice water treatment engineer to conduct high-quality jar tests; and case studies are cited illustrating how to interpret and make decisions based on the results obtained. By discussing their experiences of implementing jar tests in practice, Martin and his co-authors make this technique highly accessible even for the novice water treatment engineer. I believe that this handbook will be an essential resource for everyone working in the water quality and treatment field, from students, to research scientists, to water treatment plant operators and managers, and will certainly recommend it to my research team. I thank the authors for producing such a valuable, practical text which I am confident will be widely used.

Rita Henderson
Associate Professor
School of Chemical Engineering, UNSW Sydney

Acknowledgements

We gratefully acknowledge the indispensable cooperation, assistance, and support of all our colleagues at the Institute of Hydrodynamics of the Czech Academy of Sciences; the inspiring collaboration with other research institutions; and last but not least; the partnerships with drinking water treatment plants that help to connect the science with the practice. There are many people who contributed their experiences and knowledge to the contents of this handbook. Additionally, we thank the reviewers for their valuable input. We are also grateful to the IWA for their interest in this topic and for making this publication possible. Further, the authors acknowledge the Strategy AV21 of the Czech Academy of Sciences [VP20 – Water for Life] for valuable support.

doi: 10.2166/9781789062694_0001

Chapter 1
Introduction

Coagulation–flocculation is an important step of drinking water treatment that affects both the overall quality of treated water and the performance of downstream treatment processes, such as membrane filtration or granular activated carbon filtration. Coagulation–flocculation is capable of, at least partially, removing various common impurities, for example, turbidity causing clay minerals, algal and cyanobacterial cells, or natural organic matter (NOM) – including humic substances (HS) as well as algal organic matter (AOM). As a result of coagulation–flocculation, the impurities are trapped in aggregates formed (flocs), and are therefore removed by subsequent processes such as sedimentation, flotation, and/or deep-bed (sand) filtration.

Nevertheless, coagulation–flocculation fulfils its purpose well only if operated under convenient conditions. The key parameters are typically the type and dose of a coagulant, coagulation pH value, and mixing conditions. Improper setting of coagulation–flocculation may cause insufficient removal of target pollutants, unacceptable coagulant residuals in treated water, or formation of flocs unsuitable for the given means of floc separation. The optimal coagulation–flocculation conditions must be determined experimentally, and this is the purpose of conducting jar tests.

This book provides a brief theoretical background and comprehensive practical instructions needed for performing and evaluating jar tests. The theoretical background covers the basic principles of coagulation–flocculation, describes the influence of distinct parameters on the process efficiency and floc properties. The practical part guides the reader through the working procedure of conducting jar tests, and involves a description of the required laboratory equipment. The clues for evaluating jar test results are also provided, and some example results are discussed at length. Additionally, a manual for conducting a test of flocculation, which is aimed at the characterisation of the formed flocs, is also included.

The inspiration and motivation for preparing this publication arise from the authors' wide experience with coagulation–flocculation investigation and optimisation both in a laboratory and on a full scale at drinking water treatment plants (DWTPs). Unfortunately, some important factors affecting coagulation–flocculation are often overlooked during process optimisation, which may lead to ambiguous results. Thus, the key coagulation–flocculation parameters are emphasised herein. Additionally, coagulation–flocculation is a very complex process that involves a wide range of physicochemical aspects, and although quite a thorough theoretical background exists and some basic regularities are valid, these do not enable the optimisation of the process without conducting empirical tests. Nevertheless, the current state of the art should always be the basis for practical experiments.

doi: 10.2166/9781789062694_0003

Chapter 2
Theoretical background

This section is intended to provide only a brief summary of coagulation–flocculation principles, the most important factors governing the coagulation–flocculation efficiency, and the properties of the formed flocs. Details on the theory of coagulation–flocculation can be found elsewhere in the literature (e.g., Bratby, 2006; Gregory, 2006; Bache & Gregory, 2007). Here, we instead focus on the aspects that are often overlooked or given little attention, although they have substantial significance for the practice.

2.1 COAGULATION–FLOCCULATION

In water treatment, the term coagulation usually refers to the destabilisation of particles by dosing coagulants, and flocculation means the formation of larger aggregates – also called flocs – achieved by mixing (Bache & Gregory, 2007; Gregory, 2006). Coagulation (destabilisation) can be considered a physicochemical process because it is related to interparticle forces that follow from the chemical structures of the impurities and the coagulants (Bache & Gregory, 2007; Gregory, 2006). Flocculation (aggregation), on the other hand, is a physical process covering the relationship between the number of particles of given sizes, their collision rates and the efficiency of those collisions (Gregory & O'Melia, 1989; Gregory, 2006; Ives, 1978; Smoluchowski, 1916, 1917). However, in many cases, coagulation and flocculation are used interchangeably or together as a term describing the whole process of suspension formation in the water treatment field, and the use of the terms destabilisation and aggregation are relatively rare. Herein, the terms coagulation and flocculation are utilised in the meanings described above, while the joint term coagulation–flocculation is used to describe the whole process.

Floc formation is a result of attractive intermolecular and/or interparticle forces (electrostatic and van der Waals forces, hydrogen bonds, polymer bridges, or hydrophobic forces) that bind particles together. Nevertheless, these forces are distance-dependent and the particles need to come close enough to each other. In stable systems, this is hindered by repulsive forces (electrical double layer, steric interactions, or solvation forces) acting between particles. Therefore, to allow the particles to come closer (collide) and aggregate (form flocs), the balance in the system has to be adjusted in such a manner that attractive forces prevail. For this purpose, coagulant agents are utilised.

The other necessary condition for floc formation (aggregation) is some driving force that enables the transport of particles and therefore their contact (collision). This movement of particles can proceed via different transport mechanisms, and the process of aggregation is differentiated as follows: (i) perikinetic aggregation resulting from Brownian motion, (ii) orthokinetic – vertical aggregation, driven by gravity, and (iii) orthokinetic – horizontal aggregation, driven by the motion/flow of water.

(i) Under certain conditions, Brownian motion can result in particle collisions leading to perikinetic aggregation (Han & Lawler, 1991). In practice, perikinetic aggregation is employed only when the particles are very small, typically smaller than approximately 1 μm. Therefore, it is insufficient for the formation of flocs large enough to meet the requirements given by the water treatment technologies (Bache & Gregory, 2007; Han & Lawler, 1991).

(ii) During vertical orthokinetic aggregation (also referred to as differential sedimentation), the movement of particles is caused by gravity (Gregory, 2006; Ives, 1978). Two particles of different sizes and/or densities have different sedimentation velocities. Larger or denser particles settle faster and interact with smaller particles that have lower sedimentation velocities, which enables mutual contact and aggregation. However, vertical orthokinetic aggregation is not sufficient to achieve a collision frequency high enough to lead to satisfactory aggregation.

(iii) In practice, the water (suspension containing impurities and coagulant) is exposed to shear that is induced by mixing or flow of water, in which the transport of the particles fundamentally affects their collision rate (Camp & Stein, 1943). This process is called horizontal orthokinetic aggregation (Ives, 1978).

2.2 PARAMETERS INFLUENCING COAGULATION–FLOCCULATION EFFICIENCY

The efficiency of coagulation–flocculation is affected by several factors. In general, the most important parameters affecting the course and efficiency of coagulation are, in addition to the properties of the target impurities to be removed, coagulant type and dose and coagulation pH value. However, other parameters, such as ion content (ionic strength) or water temperature, might also be influential.

2.2.1 Influence of coagulant type and dose

The traditional and still widely used coagulants are hydrolysing iron and aluminium salts, such as ferric sulphate ($Fe_2(SO_4)_3$), ferric chloride ($FeCl_3$), or aluminium sulphate ($Al_2(SO_4)_3$) – also recognised as 'alum'. Additionally, pre-polymerised Fe- and Al-based coagulants are often employed, for example, polyferric sulphate (PFS) or polyaluminium chloride (PACl), which contain significant amount of polymeric species. Natural (e.g., alginate, chitosan, etc.) or synthetic polymers (based on polyacrylamide or polyethylenimine) can be utilised for water treatment as well, usually as flocculation aids. Furthermore, the application of coagulants comprising metals other than Fe and Al (e.g., Ti or Zr) is under investigation, and composite coagulants have also been synthesised.

As mentioned above, the most conventional coagulants are metal salts of Fe or Al that undergo hydrolysis and polymerisation upon their addition to water (see Section 2.2.2). In general, both types appeared as effective for the removal of various common impurities, such as inorganic particles, algal/cyanobacterial cells, and dissolved NOM. To illustrate, very high maximum removals of turbidity causing clay minerals (often represented by kaolin or kaolinite for research purposes) have been reported, such as approximately 95% turbidity removal when using ferric chloride (Ching *et al.*, 1994) or up to 99% turbidity removal when using aluminium sulphate (Zand & Hoveidi, 2015). Similarly, up to almost complete cell removal ($\geq$99%) was obtained under laboratory conditions when using ferric chloride (Gonzalez-Torres *et al.*, 2014), ferric sulphate (Baresova *et al.*, 2017), and aluminium sulphate (Henderson *et al.*, 2010). In general, the efficiency of dissolved NOM removal via coagulation–flocculation is usually lower than those of cells and turbidity, and there are large differences in the obtained maximum removals among the studies, owing to the specific and diverse properties of NOM. For example, the maximum removal of HS (based on dissolved organic carbon (DOC) measurements) was reported to be approximately 65–90% (Cheng, 2002; Kong *et al.*, 2021; Pivokonsky *et al.*, 2015). The maximum removal of AOM ranges approximately between 20 and 80% in terms of DOC (Baresova *et al.*, 2017, 2020; Pivokonsky *et al.*, 2009a; Zhao *et al.*, 2020). There have only been a few studies directly comparing the performance of Fe- vs. Al-based hydrolysing coagulants for certain impurities. For example, a slightly higher maximum removal of *Microcystis aeruginosa* cellular organic matter (COM) was reached by ferric sulphate (46%) than by aluminium sulphate (41%) in a study by Pivokonsky *et al.* (2009a). Similarly, coagulation with ferric sulphate resulted in a slightly better *M. aeruginosa* COM removal (45–50%) than aluminium sulphate (41–43%) in a study by Baresova *et al.* (2020). Only a small difference was also observed for *M. aeruginosa* cells, the removal of which ranged between 98.6 and 99.6% when using ferric chloride and between 95.9 and 97.6% when using aluminium sulphate under conditions applied by Gonzalez-Torres *et al.* (2014). Slightly lower residual turbidity was achieved when using ferric sulphate compared to aluminium sulphate in a kaolin removal study by Safarikova *et al.* (2013). Nevertheless, the action of hydrolysing Fe and Al salts is strongly dependent on the pH value, and the optimal coagulation pH for each coagulant may differ, which is further elaborated in Section 2.2.2. Additionally, some examples that illustrate the effect of coagulant type in practice are provided in Section 3.4.4.

Pre-polymerised Fe- and Al-based coagulants comprise a proportion of polymerised metals – polynuclear hydrolysis products prepared under certain conditions (e.g., $[Al_{13}O_4(OH)_{24}]^{4+}$ – so called 'Al$_{13}$') (Duan & Gregory, 2003). They may pose some advantages over conventional hydrolysing metal salts, such as broader optimal coagulation pH, less alkalinity consumption, or lower susceptibility to temperature effects (Jiang & Graham, 1998; Shi *et al.*, 2007; Zhang *et al.*, 2018); however, the suitability of their application always depends on the particular raw water composition. As already mentioned, the most common pre-polymerised coagulants are PFSs and PACls; however, their properties may be diverse as the procedures of their synthesis differ (Duan & Gregory, 2003; Jiang & Graham, 1998). An important characteristic of pre-polymerised coagulants is their basicity (the ratio of the moles of base added and/or bound to the moles of Al^{3+} or Fe^{3+} ions [OH]/[Al] or [OH]/[Fe]). To illustrate, PFSs of different basicities comprised diverse proportions of polymeric species and resulted in different removal efficiencies of kaolin, algal and/or cyanobacterial cells, and humic acids in a study by Lei *et al.* (2009). Similarly, PACls of different basicities displayed diverse distributions of Al species and acted differently during the coagulation of natural raw water in a study by Zhang *et al.* (2018), resulting in diverse reductions in turbidity, DOC, and UV_{254} absorbance values, as well as Al residuals. PACls with different content of Al_{13} were reported to perform differently during coagulation of certain impurities in a study by Lin *et al.* (2014). Nevertheless, pre-polymerised coagulants are generally well suited for the removal of the common impurities mentioned above, such as clay minerals, cells and NOM (Cheng, 2002; Lei *et al.*, 2009; Lin *et al.*, 2014; Zand & Hoveidi, 2015; Zhang *et al.*, 2018). While the results of some studies indicate that pre-polymerised coagulants may even outperform conventional hydrolysing coagulants comprising the same metal, some studies provide the opposite results. To illustrate, slightly higher maximum turbidity removals were obtained by using PACl (93.8–99.6%) than aluminium chloride (82.9–99.0%) in a study by Zand and Hoveidi (2015). In contrast, coagulants with preformed Al species were less effective for humic acid removal than conventional Al salt (Shi *et al.*, 2007). However, the maximum removal of humic acids was approximately 90% by using either ferric chloride of PFS in a study by Cheng (2002). Similarly, the maximum removal efficiency of non-proteinaceous fraction derived from COM of the green alga *Chlorella vulgaris* was the same (25%) when using aluminium sulphate or PACl, but aluminium sulphate performed better in terms of Al residuals (Naceradska *et al.*, 2019a). It is therefore not possible to decide whether a pre-polymerised coagulant will perform better than a hydrolysing metal salt without conducting certain experiments. Additionally, despite pre-polymerised coagulants already comprise some polynuclear metal species, their efficiency is still pH dependent (Lin *et al.*, 2014).

The coagulation efficiency often increases with the dose of a coagulant but only up to a certain value (Baresova *et al.*, 2017, 2020; Naceradska *et al.*, 2019a; Pivokonsky *et al.*, 2009a; Zhang *et al.*, 2018). In contrast, an excess of a coagulant may result in coagulation deterioration and/or high coagulant residuals in the treated water (Baresova *et al.*, 2020; Naceradska *et al.*, 2019a; Zhang *et al.*, 2018). The optimal dose of a coagulant strongly depends on the concentration and on the character of the impurities to be removed. To illustrate, Baresova *et al.* (2017) coagulated cells of the cyanobacterium *Merismopedia tenuissima* with ferric sulphate, and while the sufficient coagulant dose for a cell concentration of 4×10^5 mL^{-1} corresponded to 2 mg L^{-1} Fe, a cell concentration of 4×10^6 mL^{-1} required at least 4 mg L^{-1} Fe. In the same study, different concentrations of COM of the cyanobacterium were coagulated, that is, 3, 5, and 8 mg L^{-1} DOC, and the best removal was attained using coagulant concentrations corresponding to at least 2, 4, and 7 mg L^{-1} Fe, respectively. Ferric sulphate doses corresponding to up to 16 mg L^{-1} Fe were also tested, but the coagulation efficiency no longer improved (Baresova *et al.*, 2017). Pivokonsky *et al.* (2015) coagulated the same concentration (5 mg L^{-1} DOC) of HS and COM peptides/proteins of the cyanobacterium *M. aeruginosa* using aluminium sulphate, and while the optimal dose for HS removal corresponded to 5.5 mg L^{-1} Al, COM peptides/proteins required only 2.0 mg L^{-1} Al. Interestingly, when the compounds were coagulated together, the optimal coagulant dose was 2.8 mg L^{-1} Al, which was much lower than that for HS alone. Mutual interactions between different impurities may therefore affect the coagulant demand. Thus, the optimal dose of a coagulant must be determined experimentally for a given water composition.

2.2.2 Influence of pH value

The pH value is one of the most significant factors that influence coagulation. It affects the properties of pollutants/impurities that are to be removed from water as well as the hydrolysis of the coagulants (Duan & Gregory, 2003; Gregory, 2006; Naceradska *et al.*, 2019b; Polasek & Mutl, 1995; Wang *et al.*, 2005). Among the pH-dependent characteristics of water pollutants, the surface charge of each is essential. It governs their interactions with other particles/molecules present in water, including the coagulants, and therefore crucially affects the coagulation course and efficiency. Most inorganic particles, such as clay minerals, have a point of zero charge at a low pH and therefore carry a negative charge across a wide pH range (Bernhardt *et al.*, 1985; Safarikova *et al.*, 2013; Stumm & Morgan, 1996). The most commonly occurring organic impurities – cells of algae/cyanobacteria and dissolved NOM, which comprise humic substances and dissolved organic products of the mentioned microorganisms, that is, AOM, contain various functional groups ($-OH$, $-COOH$, $-SH$, $-NH_3^+$, $=NH_2^+$) capable of releasing or accepting a proton depending on the pH value. In general, cell suspensions, HS, and the majority of AOM carry a negative charge across a spectrum of pH values typical of natural waters (Bernhardt *et al.*, 1985; Henderson *et al.*, 2008a, 2008b; Liu *et al.*, 2009; Newcombe *et al.*, 1997; Paralkar & Edzwald, 1996; Pivokonsky *et al.*, 2015; Stumm & Morgan, 1996). However, it should be noted that certain organic substances, for example, proteins, may carry a positive charge. To illustrate, Pivokonsky *et al.* (2012) showed that the isoelectric points of peptides/proteins isolated from the cyanobacterium *M. aeruginosa* were in the pH range of 4.8–8.1. In addition to the charge, the pH value also affects the structure of some pollutants. For example, depending on the pH, some organic particles formed by macromolecular substances change their spatial structure from a macromolecular cluster to an uncoiled chain, that is, from a relatively compact particle with a small surface area to a spatial structure with a large surface area. It also affects the availability of functional groups. To illustrate, most proteins change from a compact to an expanded form at pH values <6 and >9 (Creighton, 1993).

Concerning the most frequently utilised coagulants, Fe- or Al-based salts, pH values crucially influence their hydrolysis, polymerisation, and the formation of resultant species. In brief, at low pH values (pH < 2.2 for Fe and pH < 4.5 for Al), Fe^{3+} and Al^{3+} ions occur as hexaaqua complexes [Fe(H$_2$O)$_6$]$^{3+}$ and [Al(H$_2$O)$_6$]$^{3+}$ in an aqueous medium. With increasing pH, hydrolysis proceeds to the formation of mononuclear and subsequently polynuclear hydroxo complexes, for example, [Fe$_2$(OH)$_2$]$^{4+}$, [Fe$_2$(OH)$_3$]$^{3+}$, [Fe$_3$(OH)$_4$]$^{5+}$, [Fe$_4$(OH)$_6$]$^{6+}$; [Al$_2$(OH)$_2$]$^{4+}$, [Al$_2$(OH)$_8$]$^{2-}$, [Al$_4$(OH)$_8$]$^{4+}$, [Al$_7$(OH)$_{17}$]$^{4+}$, [Al$_8$(OH)$_{20}$]$^{4+}$, [Al$_{13}$(OH)$_{32}$]$^{7+}$, [Al$_{14}$(OH)$_{32}$]$^{10+}$, [Al$_{13}$O$_4$(OH)$_{24}$]$^{4+}$ – recognised also as 'Al$_{13}$'; all comprise a certain amount of coordinated water molecules. Then, metal hydroxides, that is, Fe(OH)$_3$ and Al(OH)$_3$, are formed. They are of low solubility and can therefore form amorphous precipitates. At alkaline pH values (pH > 10.0 for Fe, pH > 8.5 for Al), anionic forms are formed. The equilibrium compositions of solutions in contact with freshly precipitated Al(OH)$_3$ and Fe(OH)$_3$ are depicted in Figure 2.1. A very simplified scheme of metal speciation following an increase

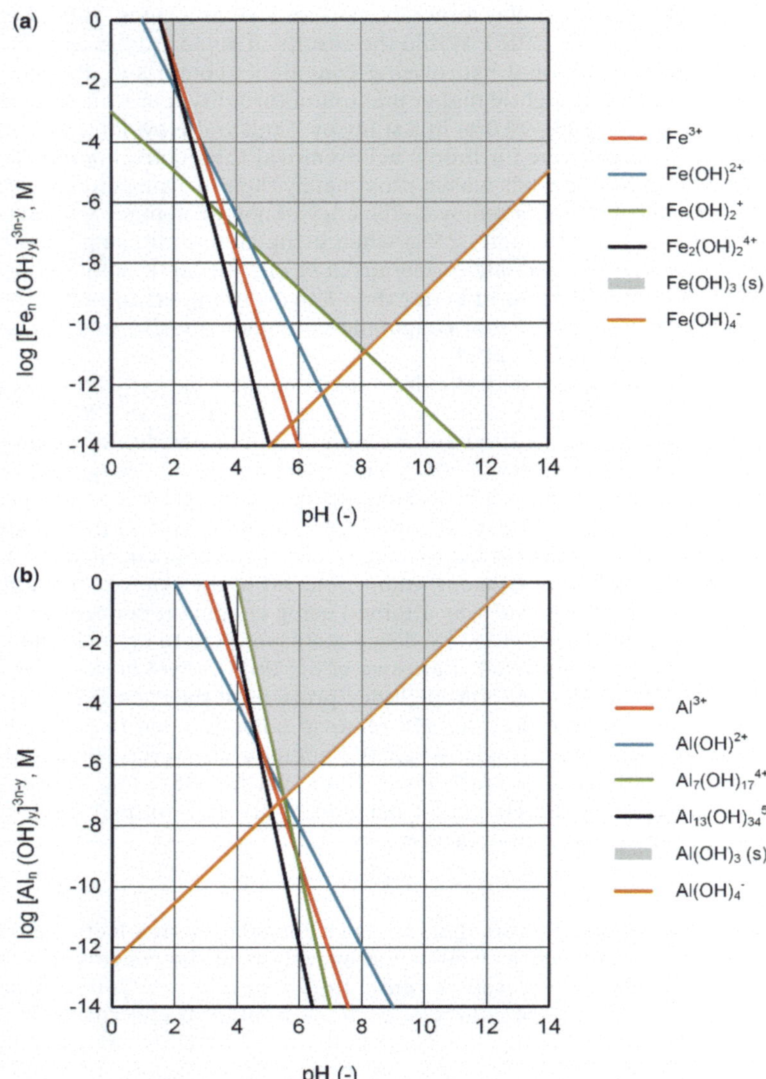

Figure 2.1 Equilibrium composition of solutions in contact with freshly precipitated (a) Al(OH)$_3$ and (b) Fe(OH)$_3$, calculated using representative values of equilibrium constants for the solubility and hydrolysis equilibria.

in the pH (omitting coordinated water molecules and the formation of polynuclear species) is as follows: Me^{3+} → Me(OH)$^{2+}$ → Me(OH)$_2^+$ → Me(OH)$_3$ → Me(OH)$_4^-$ (Duan & Gregory, 2003), while the hydrolysis of Al is linked with a much narrower pH range than that of Fe. Although pre-polymerised Fe- or Al-based coagulants (e.g., PFS or PACl) contain a significant amount of polymeric species, they still undergo hydrolysis and precipitation in parallel with the rise in pH value (Cheng, 2002; Wang *et al.*, 2004). Additionally, the charge and structure of polymeric coagulants, such as organic coagulants (e.g., alginate, chitosan, etc.) or polyacrylamide-based synthetic coagulants, also undergo alteration upon the change in pH as their functional groups accept protons or dissociate depending on the pH value.

Although the importance of coagulation pH is often overlooked, there are some studies that provide insight into the effect of pH on the coagulation of specific compounds. For example, turbidity causing inorganic impurities are usually effectively coagulated at a near neutral pH, for example, the optimal coagulation pH values for kaolin removal lie between 6.0 and 8.0 in the case of Fe-based coagulants and between 7.0 and 8.5 for Al-based coagulants. Under these conditions, kaolin is presumably removed through adsorption onto metal hydroxide precipitates (Ching *et al.*, 1994; Faust & Aly, 1998; Kim & Kang, 1998; Safarikova *et al.*, 2013). Additionally, algal/cyanobacterial cells were reported to be coagulated most effectively at pH values between approximately 6.0 and 8.0, mostly via interaction with metal hydroxides. To illustrate, Gonzalez-Torres *et al.* (2014) coagulated cells of the green alga *Chlorella vulgaris* by ferric chloride, and the maximum cell removal (up to 99.7%) was achieved at pH 6.0 and 7.0. Baresova *et al.* (2017) coagulated cells of the cyanobacterium *M. tenuissima* by ferric sulphate, and the optimum resulting in up to 99% cell removal was between pH 6.0 and 7.7. In contrast to inorganic particles and algal/cyanobacterial cells, dissolved NOM (either HS or AOM) is typically better removed under slightly acidic pH values when using hydrolysing coagulants. The optimal pH for the coagulation of HS is usually between 5.0 and 6.0 (Cheng, 2002; Liu *et al.*, 2009; Lu *et al.*, 1999; Pivokonsky *et al.*, 2015). Under these conditions, the main coagulation mechanism is assumed to be charge neutralisation between the positively charged Fe/Al hydroxopolymers and the negatively charged HS. HS removal at higher pH values ($\sim$6.0–7.0) is presumably driven by adsorption of HS onto Fe/Al precipitates, which does not result in such high HS removal efficiency (Cheng, 2002; Duan *et al.*, 2003; Liu *et al.*, 2009). The pre-polymerised Al-based coagulant PACl was also reported to perform well for HS removal at a slightly acidic pH, but it may be effective even at higher pH values. For example, 80% HS removal was obtained in the pH range of 5.0–9.0 in a study by Gao *et al.* (2006); however, sufficiently low Al residuals were reached only at pH 5.0. The pre-hydrolysed Fe-based coagulant PFS had the coagulation optimum at similar pH values as ferric chloride (Cheng, 2002). Similar to the HS coagulation, AOM is usually removed with the highest efficiency at a slightly acidic pH when using hydrolysing Fe- and Al-based coagulants. Under these circumstances, the negatively charged AOM can interact with the positively charged metal hydroxopolymers (Baresova *et al.*, 2017, 2020; Pivokonsky *et al.*, 2009a, 2009b, 2015; Widrig *et al.*, 1996). For example, Pivokonsky *et al.* (2009b) coagulated the COM of *M. aeruginosa* by using ferric sulphate and obtained the highest removals ($\sim$50%) within the pH range of 4.5–6.5. Baresova *et al.* (2017) coagulated the COM of *M. tenuissima* with ferric sulphate, and the removal was the highest ($\sim$50%) in the pH range of 5.0–6.5. Baresova *et al.* (2020) utilised either ferric or aluminium sulphate for the coagulation of *M. aeruginosa* COM, and the optimal pH values were determined to be 5.8–6.8 and 6.0–7.0, respectively. Nevertheless, when different COM components were coagulated separately, it was shown that coagulation optimums for their removal differed. Pivokonsky *et al.* (2012) coagulated the peptide–protein fraction of *M. aeruginosa* COM with ferric sulphate and obtained the maximum removal within the pH range of 4.0–6.0, which was ascribed to the charge neutralisation of the negatively charged peptides/proteins by the positively charged hydrolysis products of the coagulant. Similarly, the highest efficiency of the coagulation of *M. aeruginosa* COM peptides/proteins with aluminium sulphate was attained in the pH range of 5.2–6.7 (Pivokonsky *et al.*, 2015). In contrast, non-proteinaceous fraction of the COM had its coagulation optimum in the pH range of 7.1–7.5 when using aluminium sulphate, and in the pH range of 7.6–8.0 when using PACl; thus, this COM fraction rather interacted with the hydroxide species (Naceradska *et al.*, 2019a).

When a mixture of impurities is coagulated together (which is usually the case of natural raw water), the optimal coagulation conditions, including the pH value, might significantly differ from the optimums for single compound coagulation. For example, the presence of *M. aeruginosa* COM peptides/proteins induced the coagulation of kaolin in the pH range of 4.0–6.0 and 5.0–6.5 when using ferric and aluminium sulphate, respectively, while kaolin alone was effectively coagulated only at a near neutral pH (Safarikova *et al.*, 2013). The optimal coagulation pH for the removal of cells and the COM of *M. tenuissima* by ferric sulphate was in the range of 5.0–6.4, depending on the mutual concentration ratio, while cells alone were removed at a near neutral pH (6.0–7.7). However, in some cases, the optimal pH for a mixture of impurities may overlap with the optimums for single compounds. To illustrate, the maximum coagulation efficiency for a mixture of *M. aeruginosa* COM peptides/proteins was reached within the pH range 5.0–6.2, while the optimums for the compounds alone were pH 5.5–6.0 (HS) and 5.2–6.7 (COM peptides/proteins) (Pivokonsky *et al.*, 2015). An overview of coagulation pH values that appeared suitable for the removal of defined impurities (and their mixtures) under laboratory conditions is provided in Figure 2.2.

Nevertheless, impurities of diverse compositions and properties with different pH values suitable for their destabilisation occur in the natural waters that are to be treated by coagulation–flocculation. It is necessary to find a pH value at which the largest amount of impurities is capable of destabilisation (Polasek & Mutl, 2005; Rossini *et al.*, 1999; van Beschoten & Edzwald, 1990) to achieve the highest possible coagulation efficiency. Although theoretical knowledge on the pH dependency of destabilisation and coagulation of different impurities is very useful, the optimal coagulation pH must always be determined experimentally.

Additionally, the pH value is closely related to the acid neutralising capacity ($ANC_{4.5}$; alkalinity) value of water. The pH and $ANC_{4.5}$ values generally decrease with an increasing dose of a hydrolysing coagulant because of acid (H_3O^+ ions) release during hydrolysis. To illustrate, the reactions of 1 mg L^{-1} of ferric sulphate ($Fe_2(SO_4)_3 \cdot 9H_2O$), ferric chloride ($FeCl_3 \cdot 6H_2O$), and aluminium sulphate ($Al_2(SO_4)_3 \cdot 18H_2O$) to form the corresponding hydroxide precipitates require 0.0107, 0.0110, and 0.0090 mmol L^{-1} of alkalinity, respectively (Naceradska *et al.*, 2019b). Also pre-polymerised coagulants affect the pH and $ANC_{4.5}$ to a certain degree. Therefore, in many cases, it is often necessary to pretreat raw water by adding acid or base to achieve the optimal coagulation pH value.

2.2.3 Influence of other parameters

Temperature is another factor influencing the coagulation–flocculation process. One reason is that it affects the hydrolysis of coagulants (Xiao *et al.*, 2008). Temperature appeared to influence the pH of the minimum solubility of diverse coagulants, including hydrolysing metal salts and various pre-polymerised coagulants (Pernitsky & Edzwald, 2003). As a result, the temperature may alter the optimal coagulation pH (Bache & Gregory, 2007). Additionally, very low temperatures were reported to disturb coagulation–flocculation (Morris & Knocke, 1984). The adverse effect of low temperature was reported to be more severe for Al-based coagulants than for Fe-based hydrolysing coagulants (Morris & Knocke, 1984; Wang *et al.*, 2005). In this regard, the application of pre-polymerised coagulants might also be beneficial in the case of low water temperature (Bache & Gregory, 2007). Furthermore, the temperature affects the viscosity of water, which may be manifested on the flocculation timescale, although the influence of water chemistry is considered more important (Bache & Gregory, 2007; Fitzpatrick *et al.*, 2004).

The course of coagulation can also be influenced by the presence and concentrations of ions or the ionic strength in general. It affects, for example, the destabilisation of impurities (Gregory, 2006) or the solubility of salts (Wang *et al.*, 2005), and so on.; and some ions may be involved in coagulant precipitation (Gregory, 2006).

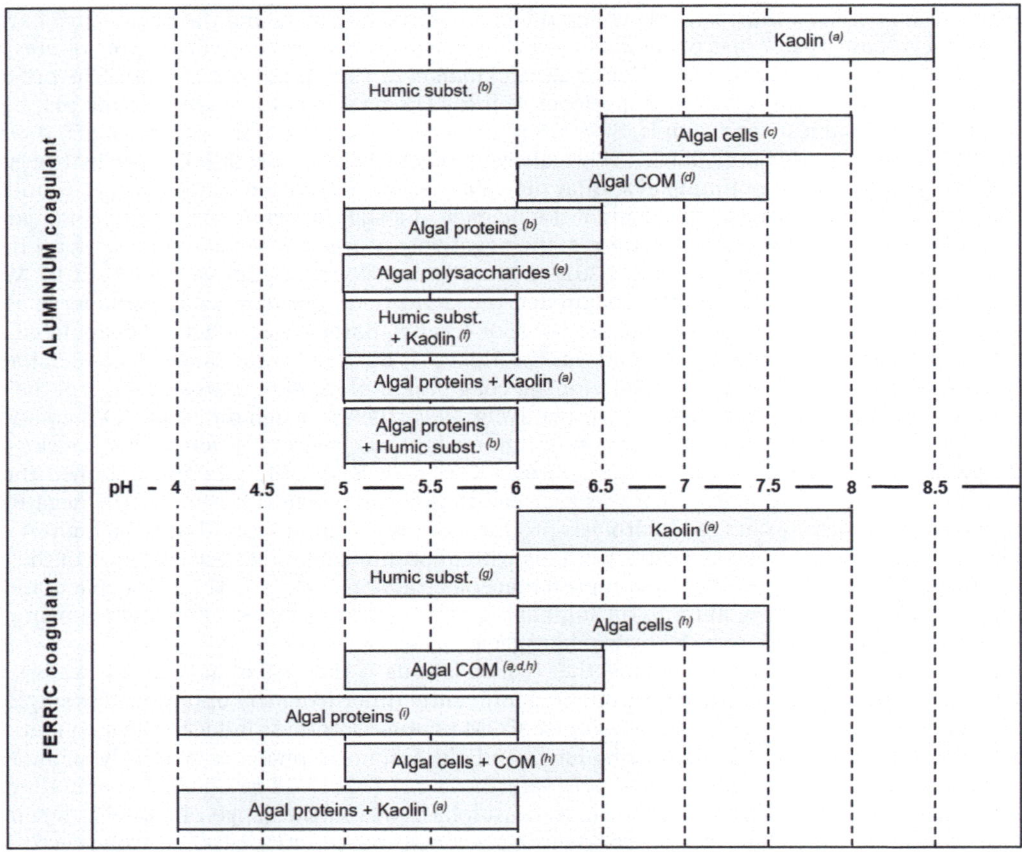

Figure 2.2 Overview of the coagulation pH values suitable for the removal of specific impurities and their mixtures when using hydrolysing coagulants (adapted from Naceradska *et al.*, 2019b). References: [a]Safarikova *et al.* (2013); [b]Pivokonsky *et al.* (2015); [c]Henderson *et al.* (2008a); [d]Baresova *et al.* (2020); [e]Naceradska *et al.* (2019a); [f]Li *et al.* (2014); [g]Duan *et al.* (2003); [h]Baresova *et al.* (2017); [i]Pivokonsky *et al.* (2012).

In general, where as the coagulant type and dose and the coagulation pH value are adjustable coagulation parameters, the temperature and the composition (therefore also the ionic strength and concentration of distinct ions) of water to be treated are given. Owing to the focus of this book, a detailed elaboration of the effect of the unalterable parameters is therefore avoided.

SUMMARISING BOX

Both coagulant type/dose and coagulation pH value are critical parameters affecting the course and efficiency of coagulation–flocculation. It must be taken into consideration that the addition of most coagulants significantly affects the pH, and therefore adjusting the pH is often necessary. In general, the optimal coagulant type/dose and the optimal coagulation pH value depend on many factors and they need to be determined experimentally.

2.3 PARAMETERS INFLUENCING FLOC PROPERTIES AND THEIR SEPARATION EFFICIENCY

As previously mentioned, the formation of flocs/aggregates is the result of the action of attractive and tangential (shear) forces on the coagulation participants. It follows that the resulting floc properties are dependent on the factors that determine these forces. Floc properties are, therefore, affected not only by the composition and concentrations of their components (impurities and coagulants), but also by mixing parameters, that is, the mixing intensity (expressed by the global shear rate/mean velocity gradient), the distribution of the tangential forces in the mixed volume, and the mixing time (see the scheme in Figure 2.3). The floc properties of interest include their size distribution, structure, shape, density, and so on.

The amounts and characteristics of impurities, such as the character and number of functional groups, their molecular weights, charges, and so on., together with the doses and properties of the utilised coagulants, determine the intermolecular interactions (electrostatic, van der Waals, hydrophobic, hydrogen bonding, etc.) that hold individual floc components together (Bache & Gregory, 2007; Birdi, 2003; Hunter, 2001). Accordingly, they also

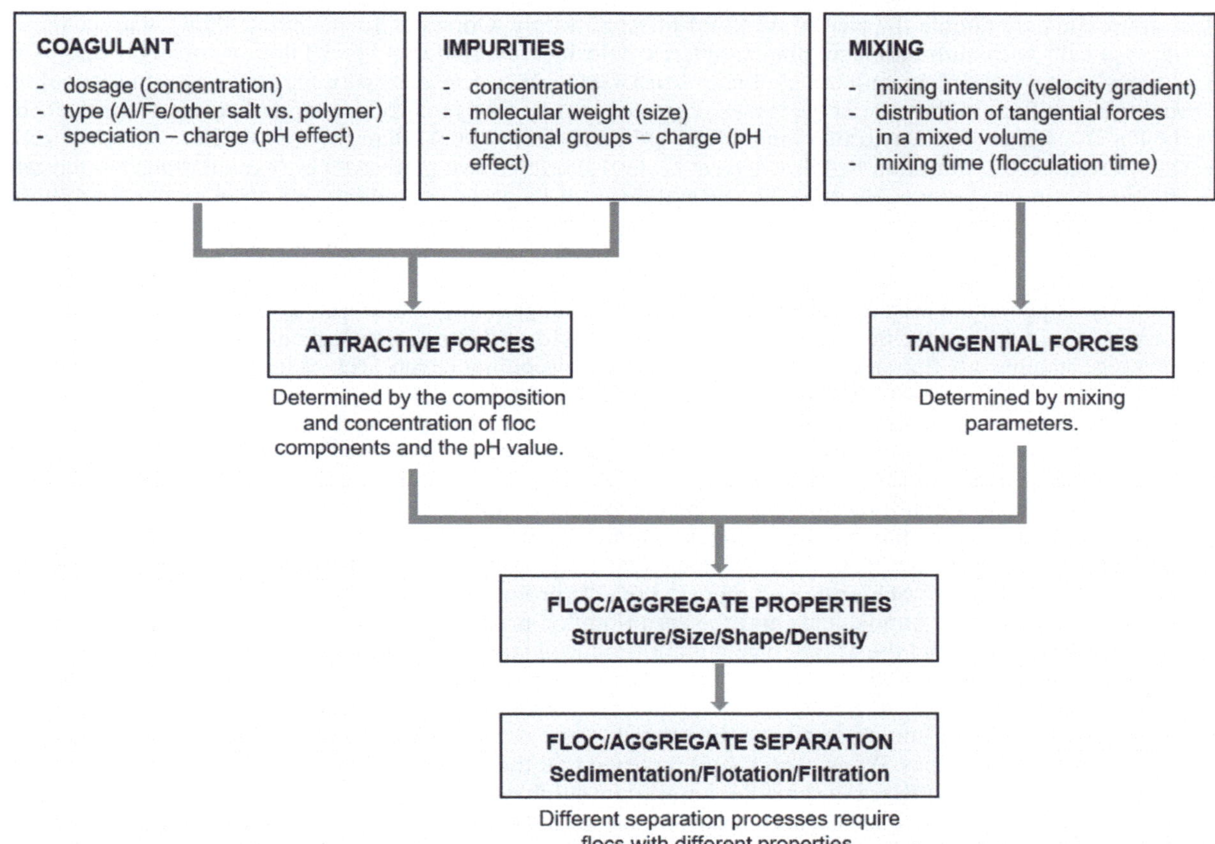

Figure 2.3 Schematic illustration of the factors influencing floc properties.

affect the floc properties. Additionally, the effect of pH value is also considerable since it determines the magnitude and distribution of the charges of the impurities as well as the speciation and charges of the coagulants. In practice, the raw water composition is given; therefore, only the type and dose of a coagulant and the coagulation pH value can be chosen and adjusted. Providing that optimal coagulation conditions are set, floc properties are further influenced primarily by the mixing intensity (see Section 2.3.3) (Bache & Rasool, 2001; Bache *et al.*, 1999; Bouyer *et al.*, 2005; Bubakova *et al.*, 2013; Coufort *et al.*, 2005, 2007; Filipenska *et al.*, 2019; Francois, 1987; Gorczyca & Ganczarczyk, 1999; Li *et al.*, 2006; Mutl *et al.*, 2006; Pivokonsky *et al.*, 2011; Serra *et al.*, 2008; Spicer *et al.*, 1998; Xu *et al.*, 2010). The other significant factors influencing mainly the floc size distribution are the mixing time (see Section 2.3.4) and the velocity gradient (shear rate) distribution within the mixed volume (Bubakova *et al.*, 2013; Coufort *et al.*, 2005, 2007; Filipenska *et al.*, 2019; Mutl *et al.*, 2006). While the mixing intensity and time can be adjusted to prepare flocs with properties suitable for the required type of separation, the need for the homogeneous distribution of the tangential forces is hard to achieve in practice. In most of the technological mixing/flocculation devices (paddle mixers, baffled chambers, etc.) the tangential force distribution in the mixed volume is uneven. As a result, a heterogeneous suspension with a broad floc size distribution can be formed. A relatively homogeneous velocity gradient distribution can be achieved using the fluidised layer of a granular material in full-scale operation (Mutl *et al.*, 1999; Pivokonsky *et al.*, 2008) or the Taylor–Couette reactor in a laboratory as mixing devices (DiPrima *et al.*, 1984).

The properties of flocs determine their separation efficiency. For instance, large and dense flocs are needed for sedimentation, small and strong flocs for sand filtration, and small flocs easily attachable to air bubbles for dissolved air flotation (Vasatova *et al.*, 2020). The influence of the floc properties on their separation is described in detail in Section 2.3.5.

2.3.1 Effect of coagulant type on floc properties

Apparently, there is a relationship between the floc properties (particularly their size) and the metal comprised as the main component of a coagulant. For instance, Gregory and Yukselen (2001) coagulated tap water enriched with kaolin and humic acids by four different coagulants: aluminium sulphate, PAXXL9 (a PACl of a given basicity), ferric sulphate, and PIX115 (an Fe-based pre-polymerised coagulant). The size of the resulting flocs decreased in the order PIX115 > PAXXL9 > ferric sulphate > aluminium sulphate. For both Al- and Fe-based coagulants, the pre-polymerised product gave significantly increased floc size. Jarvis *et al.* (2005a) reported that Fe-NOM flocs reached approximately double the size of Al-NOM flocs. Similarly, Gonzalez-Torres *et al.* (2014), who coagulated *M. aeruginosa* cells with aluminium sulphate and ferric chloride, revealed that Fe-cell flocs were larger than Al-cell flocs. Filipenska *et al.* (2019) coagulated kaolin and/or COM of *M. aeruginosa* with aluminium and ferric sulphate and revealed larger flocs in the case of Fe. It was suggested that Fe tends to produce larger hydrolysis products than Al, as the Fe–O distance in the hydrated ion $[Fe(H_2O)_6]^{3+}$ is greater (2.00 Å) than the Al–O distance in $[Al(H_2O)_6]^{3+}$ (1.89 Å) (Persson, 2010). Additionally, Filipenska *et al.* (2019) offered two more possible explanations related rather to coagulation optimisation, namely, that the optimal doses of Fe-based coagulants are often higher than those of Al-based coagulants (therefore producing larger amounts of precipitates) or that the biopolymers contained in AOM may have a spatially larger arrangement at a more acidic pH, which is more favourable for Fe-based coagulation than for Al-based coagulation.

Additionally, coagulants based on Zr and Ti are under investigation. For example, Jarvis *et al.* (2012) compared the characteristics of flocs made by the coagulation of NOM with different coagulants under optimised conditions, and the average steady-state floc sizes decreased in the order zirconium oxychloride > ferric sulphate > aluminium sulphate. Furthermore, several studies reported that flocs formed by using titanium tetrachloride were larger than those formed by using other coagulants. To illustrate, Zhao *et al.* (2011a) found that the size of NOM comprising flocs at the steady state was higher when using titanium tetrachloride rather than PACl. Zhao *et al.* (2011b) concluded that the size of humic acid and kaolin comprising flocs at the steady state followed the order titanium tetrachloride > ferric chloride > PFS > aluminium sulphate > PACl. Zhao *et al.* (2013a) coagulated river water containing NOM, and the floc size decreased depending on the utilised coagulant in the order titanium tetrachloride > ferric chloride > aluminium sulphate. Hussain *et al.* (2014) coagulated NOM-comprising raw water by aluminium sulphate, zirconium tetrachloride, and titanium tetrachloride and reported that Ti produced the largest flocs. Zhao *et al.* (2013b, 2015) compared the coagulation of humic acids and the resulting floc properties when using polytitanium chloride and PACl of different basicities, and the Ti-based coagulant produced larger flocs than the Al-based coagulants.

Some studies reported smaller floc sizes when using pre-polymerised coagulants rather than hydrolysing coagulants. For example, Shi *et al.* (2007) and Wang *et al.* (2009) suggested that the smaller size of flocs formed by PACl compared to aluminium chloride may be related to the reconformation of the long humic acid molecules around the preformed Al species. Wang *et al.* (2016) reported that the size of fulvic acid comprising flocs prepared by using different coagulants decreased in the order aluminium sulphate > PACl > Al_{13} polymer, which was ascribed to the different coagulation mechanisms. Additionally, Jiao *et al.* (2016), who used aluminium sulphate and nano-Al_{13} for the formation of Al-NOM flocs, reported a larger floc size for aluminium sulphate. In contrast, Gregory and Dupont (2001) reported that commercial PACl products of various basicities (PAX-16, PAX-XL1, and PAX-XL19) formed larger flocs than aluminium sulphate – the higher the basicity, the larger the flocs. The larger floc size when using pre-polymerised coagulants compared to hydrolysing coagulants (both Al- and Fe-based) was also found by Gregory and Yukselen (2001), as mentioned above.

Nevertheless, the comparison of the coagulant effect on the floc properties (size) has to be made very cautiously due to the very different experimental conditions applied in published studies. The coagulation pH value was not controlled in many of the studies, which might have distorted the results. However, some general conclusions concerning floc size can be drawn. For the coagulant type, Al-based coagulants form smaller flocs than Fe-based coagulants (Filipenska *et al.*, 2019; Gonzalez-Torres *et al.*, 2014; Jarvis *et al.*, 2005a, 2012; Zhao *et al.*, 2013a), and Zr (Jarvis *et al.*, 2012) and Ti salts result in the formation of the largest flocs (Chekli *et al.*, 2017a, 2017b; Zhao *et al.*, 2013a). If the idea suggested by Filipenska *et al.* (2019) that the floc size increases with increasing M–O distance in a hydrated metal ion is accepted, then the order of floc size Ti > Zr > Fe > Al would correspond to M–O distances of 2.34 > 2.19 > 2.00 > 1.89 Å, respectively, as reported by Persson (2010). Pre-polymerised metal salts, particularly PACls of different Al species, prepared in a laboratory were mostly reported to form smaller flocs than their monomeric versions (Jiao *et al.*, 2016; Shi *et al.*, 2007; Wang *et al.*, 2009, 2016). However, commercially available pre-polymerised products (e.g., PAX-16, PAX-XL1, PAX-XL9, or PIX115) formed larger flocs than their corresponding metal salts (Gregory & Dupont, 2001; Gregory & Yukselen, 2001). The reason for this phenomenon remains unclear. The concentration of the coagulant affects the floc size as well. Most researchers (Barbot *et al.*, 2010; Gonzalez-Torres *et al.*, 2014; Pivokonsky *et al.*, 2009a) agree that the higher the coagulant concentration is, the larger the flocs.

2.3.2 Effect of raw water composition on floc properties

Knowledge of the effect of raw water composition on the floc size or their other properties is limited. Nevertheless, smaller flocs are apparently formed during coagulation of clay minerals (e.g., kaolinite) than NOM represented by either HS or AOM when using hydrolysing coagulants (Filipenska *et al.*, 2019; Henderson *et al.*, 2006). The reason is probably the polymeric character (high molecular weight) of some NOM components, which allows flocs to grow to significant sizes (Filipenska *et al.*, 2019; Vandamme *et al.*, 2014). To illustrate, the floc sizes determined by Filipenska *et al.* (2019) decreased depending on the coagulated impurity composition and on the coagulant in the order of COM + kaolinite + Fe > COM + Fe > COM + kaolinite + Al > COM + Al > kaolinite + Fe > kaolinite + Al. Additionally, floc size was found to be dependent on the AOM composition, particularly on the distribution of proteins and carbohydrates. Thus, the coagulation of AOM produced by different species may result in flocs of diverse properties (Gonzalez-Torres *et al.*, 2017).

2.3.3 Effect of mixing intensity on floc properties

Mixing intensity expressed by the mean velocity gradient (or global shear rate) – $\bar{G}$ – is one of most important factors influencing floc properties; the 'mean' velocity gradient is usually referred to since the velocity gradient within the mixed volume is not uniform under real conditions. The impact of $\bar{G}$ on the floc size distribution is illustrated in Figures 2.4 and 2.5. Generally, both the average and maximum floc size decrease with increasing $\bar{G}$ (Bache *et al.*, 1999; Bouyer *et al.*, 2004; Bubakova *et al.*, 2013; Filipenska *et al.*, 2019; Francois, 1987; Parker *et al.*, 1972; Wang *et al.*, 2017). At $\bar{G}$ values lower than 100–150 s^{-1}, the decrease is very steep, and even a small increase in $\bar{G}$ results in a large change in the floc size. At $\bar{G}$ values greater than 100–150 s^{-1}, the decrease in floc size with $\bar{G}$ is much slower (Figure 2.4(a)). Additionally, with increasing $\bar{G}$, the flocs are more homogenous in size. Above certain $\bar{G}$ values, the floc size distribution curves are almost identical (Figure 2.4(b)).

The dependence of the floc size on $\bar{G}$ is described by the following equation (Parker *et al.*, 1972; Bouyer *et al.*, 2004; Jarvis *et al.*, 2005b):

$$d_{\mathrm{av/max}} = C\bar{G}^{-2\gamma}, \tag{2.1}$$

where $d_{\mathrm{av/max}}$ is the average/maximum diameter of flocs in the system (simply the floc size), $\bar{G}$ is the mean velocity gradient (global shear rate) and C and γ are constants that are influenced by the composition and concentration of contaminants, type and dose of a coagulant, and pH value. The exponent γ, however, does not have a constant value throughout different $\bar{G}$ values (Parker *et al.*, 1972; Tambo & Watanabe, 1979). Figure 2.5 illustrates the $\bar{G} - d$ dependence (as measured by different researchers) on a log–log scale, where the slope changes of the different lines

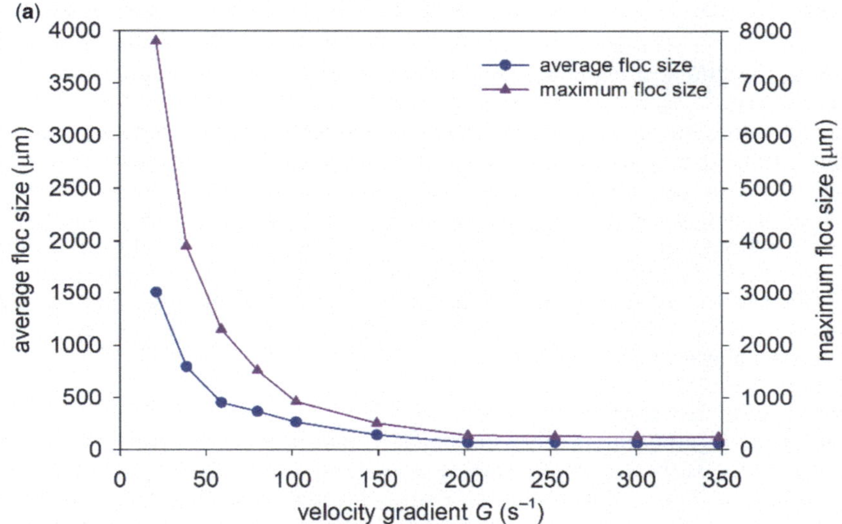

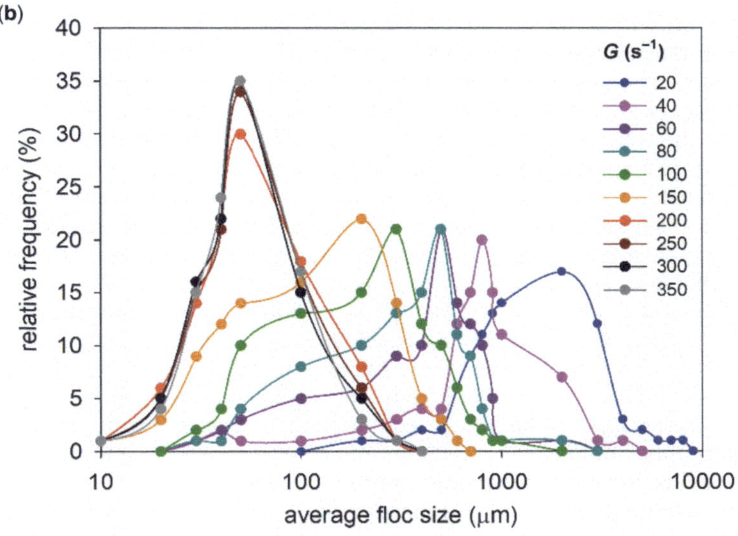

Figure 2.4 Effect of the velocity gradient on (a) the average and maximum floc sizes and (b) the floc size distribution (adapted from Bubakova *et al.*, 2013).

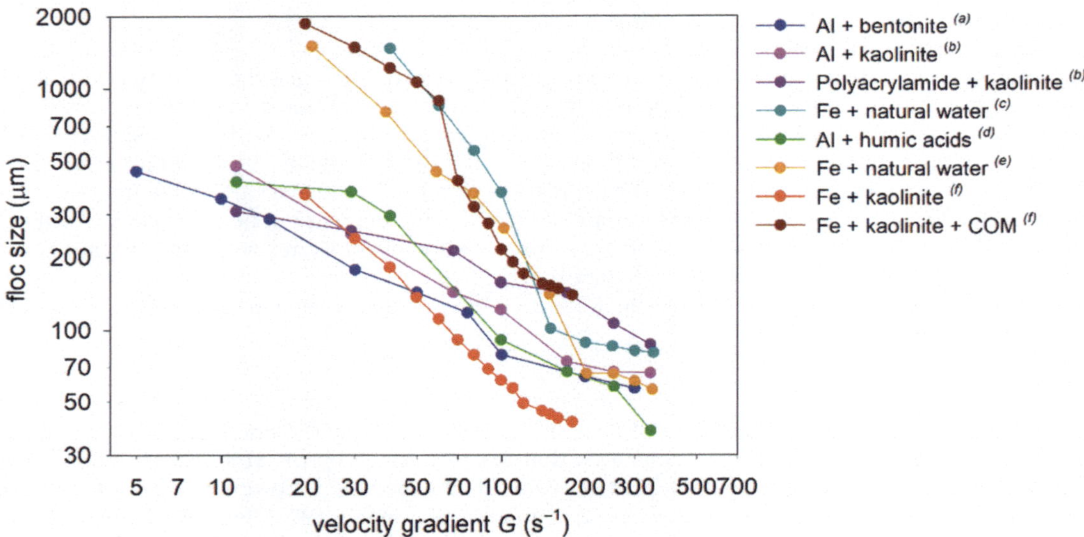

Figure 2.5 Influence of the velocity gradient on the size of the formed flocs according to different studies. References: [a]Bouyer *et al.* (2004); [b]Li *et al.* (2006); [c]Mutl *et al.* (2006); [d]Wang *et al.* (2009); [e]Bubakova *et al.* (2013); [f]Filipenska *et al.* (2019).

are clearly visible. Due to the different compositions of the tested waters and coagulants used, the values of $\bar{G}$ at which this change occurs vary.

In practice, different $\bar{G}$ values may be applied stepwise. In general, flocculation mixing can be divided into rapid mixing ($\bar{G}$ higher than ~100 s⁻¹) and slow mixing ($\bar{G}$ lower than ~100 s⁻¹). Nevertheless, the boundary between low and high $\bar{G}$ is not fixed. Additionally, flocculation mixing should be preceded by homogenisation mixing at a $\bar{G}$ value higher than approximately 300 s⁻¹, the purpose of which is to efficiently disperse the coagulant (Vasatova *et al.*, 2020). Appropriate adjustment of $\bar{G}$ and therefore the size and other properties of the flocs may be utilised to achieve the maximum efficiency of the subsequent separation step (details in Section 2.3.5). However, the floc size (and generally all floc properties) is influenced not only by $\bar{G}$ but also by other factors. Therefore, the mixing intensities have to be determined individually for a given water composition and coagulation conditions applied with respect to the following aggregate separation.

2.3.4 Effect of mixing time on floc properties

Additionally, the mixing time significantly affects the floc size distribution (Bouyer *et al.*, 2005; Bubakova *et al.*, 2013; Coufort *et al.*, 2005, 2007; Mutl *et al.*, 2006). During coagulation–flocculation, 2–3 phases of floc size evolution accompanied by changes in the number of flocs and their structure can be observed (Bubakova *et al.*, 2013; Moruzzi *et al.*, 2017). These are (i) floc growth, (ii) floc break-up and/or restructuralisation (which does not take place in some cases), and (iii) steady state (see Figure 2.6).

(i) In the first phase – floc growth – the size of flocs typically grows very rapidly and steeply with time. The relationship can be linear (Heath & Koh, 2003; Nan *et al.*, 2016; Prat & Ducoste, 2006; Selomulya *et al.*, 2003; Spicer *et al.*, 1996; Williams *et al.*, 1992) or exponential (De Boer *et al.*, 1989; Kusters *et al.*, 1997; Oles, 1992; Selomulya *et al.*, 2003). Ehrl *et al.* (2009) assumed that this difference can be caused by different sizes of primary particles that determine the relative significance of perikinetic to orthokinetic aggregation. Another reason can be the effect of the coagulation conditions, that is, pH value (see Figure 2.7), or possibly coagulant dose (Cao *et al.*, 2011). Experiments showed that the lower the $\bar{G}$ value was, the steeper the floc growth rate. This is caused by the higher collision efficiency at low $\bar{G}$ (Francois, 1987; Oles, 1992; Selomulya *et al.*, 2001). Floc growth is logically accompanied by a decrease in the floc number in the system. Simultaneously, the porosity and irregularity of flocs increase. The maximum size peak is noticeable mainly when low $\bar{G}$ is applied. This peak becomes less sharp with increasing $\bar{G}$, and at very high $\bar{G}$ (> 250 s⁻¹), it is practically non-existent (Figure 2.6). The reason is that small tangential forces at low $\bar{G}$ enable flocs to grow to large sizes (but it is noteworthy that the flocs can easily break-up or restructuralise due to the non-homogeneity of the velocity gradient field), while the high tangential forces at high $\bar{G}$ prevent flocs from growing to large sizes in the initial flocculation phase.

(ii) The second phase (if present) is called floc break-up and/or restructuralisation. A floc size decrease accompanied by a floc number increase is characteristic of this phase, indicating that large flocs break up due to the non-homogeneous velocity gradient distribution (Bubakova *et al.* 2013; Hopkins & Ducoste, 2003; Prat & Ducoste, 2006; Selomulya *et al.* 2001; Spicer *et al.* 1996; Williams *et al.* 1992). If the floc number does not change significantly with decreasing floc size, restructuralisation and compaction of flocs occur (Selomulya *et al.*, 2003). As previously mentioned, some studies have not reported the presence of this phase. However, this may be due to the premature termination of the tests (Cao *et al.*, 2011; He *et al.*, 2018a, 2018b, 2019; Prat & Ducoste, 2006).

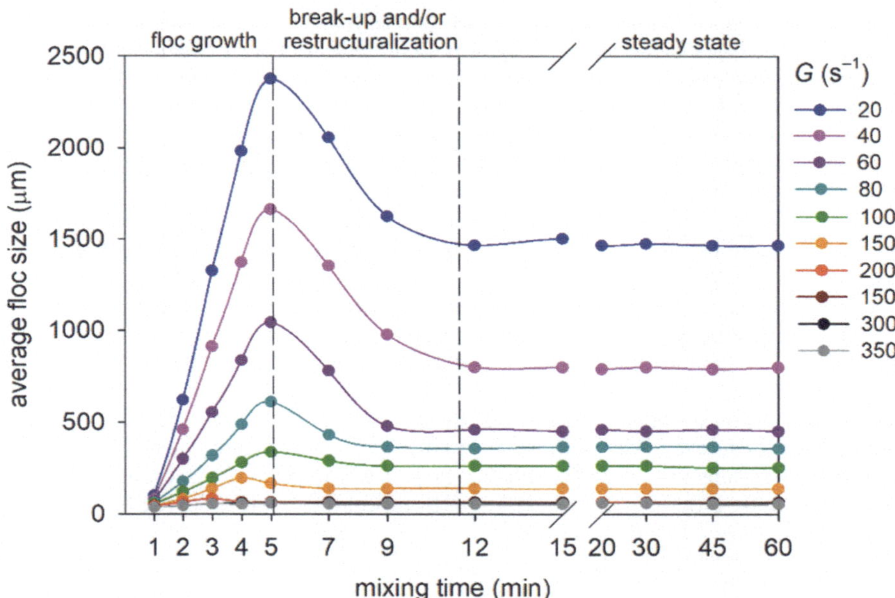

Figure 2.6 Average floc size evolution over time under different velocity gradients (adapted from Bubakova *et al.*, 2013).

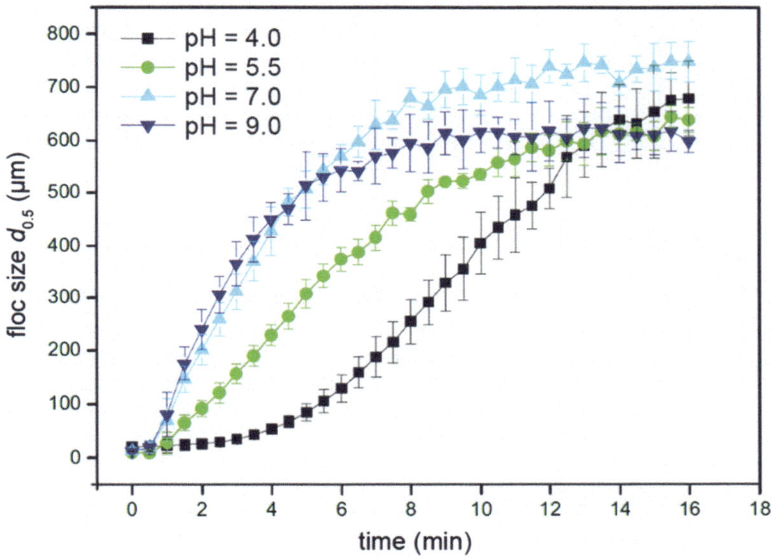

Figure 2.7 Differences in the floc size evolution over time depending on the pH value (adapted from Cao *et al.*, 2011).

(iii) After a definite mixing time, the floc size (and other properties) does not change any further – thus a steady state is reached. Concerning the local distribution of the tangential forces, break-up and restructuralisation can still take place, but these processes are generally balanced (Oles, 1992; Spicer & Pratsinis, 1996a, 1996b). The time needed to achieve a steady state decreases with the applied velocity gradient (Oles, 1992; Spicer *et al.*, 1996). It is usually in the range between several minutes to several tens of minutes (Bubakova *et al.*, 2013; Mutl *et al.*, 2006).

2.3.5 Influence of floc properties on their separation

Floc properties such as size distribution, density, and therefore settling velocity, which can be controlled by adjusting the mixing intensity and time (as described in Sections 2.3.3 and 2.3.4), significantly influence the efficiency of floc separation. In general, two types of aggregate separation can be distinguished: single-stage (i) and double-stage (ii, iii) separation, the suitability of which depends on both the amount and properties of the flocs (Vasatova *et al.*, 2020).

(i) Single-stage separation by sand filtration
Direct filtration requires flocs that penetrate the entire volume of a filter bed and exhibit sufficient ability to adhere to the filter cartridge surface (sand). Small flocs, generally <50–60 µm with high density, have been

proven to be ideal for direct filtration separation (Bache & Gregory, 2010; Ngo *et al.*, 1995; Pivokonsky *et al.*, 2012). In contrast, large flocs (>100 μm) of irregular shape and relatively low density are not suitable for sand filtration because they are not able to penetrate deep into the filter bed. They clog the sand filter, which causes a rapid pressure drop, and excessive filter cycles (because the filter run times are shortened) (Bubakova & Pivokonsky, 2012; Pivokonsky *et al.*, 2011). The velocity gradients and residence (mixing) time required to form suspensions well suited for direct filtration roughly correspond to the rapid flocculation mixing values ($\bar{G} = 100\text{–}400$ s^{-1}) (Pivokonsky *et al.*, 2011; Polasek, 2011).

(ii) Double-stage separation by sedimentation and sand filtration

For separation by sedimentation, large flocs (>100 μm) with high density and resistance to tangential forces are ideal (Edzwald, 1995; Polasek & Mutl, 2005). Smaller flocs eventually remain in water after sedimentation and then continue to the filtration step. It has been suggested (Polasek, 2011) and confirmed at several DWTPs that large flocs with high density can be prepared by a sequence of rapid mixing followed by slow flocculation mixing. The formation of large flocs can be supported by applying a flocculation aid at the end of the rapid mixing phase.

(iii) Double-stage separation by flotation and sand filtration

During dissolved air flotation, flocs adhere to rising air bubbles that carry them to the water surface. Therefore, small flocs of tens of μm, preferably 25–50 μm, with low density (close to the density of water) are required (Edzwald, 2010). However, highly efficient flotation has been reported for even larger (>50 μm) low-density flocs formed by low velocity gradients ($\bar{G} = 10$ s^{-1}), possibly because the flocs have been broken into sizes suitable for flotation as they enter into the contact zone where $\bar{G}$ values are 10–100 times higher than those used for floc formation (Bache & Gregory, 2010). Nevertheless, since flotation generally requires only very small flocs, the residence (mixing) time in flocculation should usually be up to 10 min (Edzwald, 2010; Valade *et al.*, 1996). Thus, flocculation does not reach a steady state and usually continues in the flotation unit.

The efficiencies of flotation and sedimentation are sometimes compared. Flotation is typically reported to be more effective than sedimentation when treating water containing cyanobacteria, algae, humic substances, or low amounts of inorganic turbidity forming particles. The low density of the resulting flocs is usually stated as the reason for the high flotation efficiency (Edzwald, 2010). However, this explanation is based on the erroneous assumption that the separations of flocs formed under the same mixing conditions were compared (Khiadani *et al.*, 2013). The ranges of $\bar{G}$ values recommended for the different mixing stages in relation to subsequent types of separation are summarised in Table 2.1.

Table 2.1 Recommended mixing conditions for the formation of flocs suitable for the different types of separation.

Type of separation	Recommended floc size for this type of separation	Mixing type, its purpose and recommended velocity gradients ($\bar{G}$)*		
		Homogenisation (applied after coagulant dosing)purpose: achieving a dispersion of the coagulant in water	**Rapid flocculation** (applied after homo-genisation)purpose: producing small and dense flocs	**Slow flocculation** (applied after rapid flocculation)purpose: producing large flocs (dense for sedimentation, floating for flotation)
Filtration	50–60 μm[a,b] (max 100 μm)	$\bar{G} > 300$ s^{-1c}	$\bar{G} = 100\text{–}400$ s$^{-1d,e}$	–
Sedimentation + filtration	>100 μm[f,g]	$\bar{G} > 300$ s^{-1c}	$\bar{G} = 100\text{–}400$ s$^{-1d,e}$	$\bar{G} = 20\text{–}100$ s^{-1e}
Flotation + filtration	25–50 μm[f]	$\bar{G} > 300$ s^{-1c}	$\bar{G} = 50\text{–}100$ s^{-1f} (sometimes 150 s^{-1})	$\bar{G} = 50\text{–}100$ s^{-1f} (sometimes 30 s^{-1})

References: [a]Bubakova and Pivokonsky (2012); [b]Ngo *et al.* (1995); [c]Morrow and Rausch (1974); [d]Pivokonsky *et al.* (2011); [e]Polasek (2011); [f]Edzwald (2010); [g]Edzwald (1995).

*These are ranges of $\bar{G}$ values since the floc properties (size, density, etc.) are influenced also by the type and concentration of coagulants and impurities (Filipenska *et al.*, 2019). For example, flocs prepared with Fe-based coagulants are larger than those prepared with Al-based coagulants, or AOM flocs are larger than kaolinite flocs; thus, lower $\bar{G}$ values should be applied in the case of an Fe coagulant or AOM flocs, respectively, to obtain flocs of similar size.

SUMMARISING BOX

Both mixing intensity and mixing time significantly affect the properties of flocs arising as a result of coagulation–flocculation. In practice, a stepwise application of different $\bar{G}$ values is often appropriate. The floc properties are of great importance with regard to their subsequent separation, for example, different floc characteristics are more suitable for sedimentation than for sand filtration.

doi: 10.2166/9781789062694_0015

Chapter 3

Jar tests

3.1 PRINCIPLE OF JAR TESTS

Jar tests performed under laboratory conditions model the floc formation during water treatment by coagulation–flocculation. In addition to research purposes, jar tests are essential for the evaluation and optimisation of operating parameters of existing water treatment plants and to determine the design parameters of new plants.

For example, jar tests with the following objectives can be performed based on the specific requirements:

1. Optimisation of the **coagulant dose** and **coagulation pH value** under the given mixing conditions (corresponding to the proposed or existing technology). Notably, coagulant addition affects the pH; thus, optimisation of these parameters should be done jointly.
2. Optimisation of the **rapid flocculation mixing conditions** (velocity gradient and mixing time) for the conditions determined according to point 1.
3. Optimisation of the **slow flocculation mixing conditions** (velocity gradient and mixing time) for the conditions determined in point 2.
4. **Evaluation and control of the doses of the coagulant and other reagents** under the optimised rapid and slow flocculation mixing conditions determined in points 2 and 3.
5. Optimisation of the **flocculation aid dose** under the conditions specified in the previous points.

The jar test procedure is modified according to the purpose of optimisation and the parameters that need to be optimised. Jar tests following the full range of points 1–5 shown above are usually performed only when new technologies for future water treatment plants or reconstructions of existing treatment plants are being considered. In the common practice of water treatment plant operation, jar tests are performed only to a limited extent (usually only point 1) to optimise and control adjustable operating parameters (coagulant dose and coagulation pH value).

The optimisation of the coagulation–flocculation operating parameters must be performed every time any change of raw water quality occurs, for example, a change of pH; $ANC_{4.5}$; turbidity value; concentration of organic substances – total/dissolved organic carbon (TOC/DOC), chemical oxygen demand by potassium permanganate (COD_{Mn}), ultra-violet absorbance at the wavelength of 254 nm (UV_{254}); manganese concentration; optical density; number of microorganism cells; and so on. In these cases, only the dose of a coagulant and the coagulation pH value are typically the subjects of laboratory optimisation. However, in this situation, it is necessary to ensure that the optimisation jar tests will consider the whole water treatment scheme at the given facility. For example, if there is a double-stage separation of the floc suspension by sedimentation and filtration at the treatment plant, it is necessary to appropriately simulate the slow flocculation mixing with the required mixing time to prepare suitable aggregates in laboratory jar tests and subsequently to simulate sedimentation and filtration of the resulting floc suspension (Pivokonsky *et al.*, 2011). In general, the mixing conditions applied during the jar tests must correspond to the actual conditions at the treatment plant. It is particularly important to apply the times of rapid mixing and slow flocculation mixing that correspond to the real situation. If the real values of the mixing intensities (velocity gradients) used at the facility are not known, they must be determined (measured). The values of velocity gradients actually used in practice must then be transformed into values corresponding to the stirrers used in the jar test (see Section 3.3).

However, if for some reason it is not possible to determine the velocity gradient values applied at a water treatment plant, the following mixing sequence and velocity gradients are recommended for laboratory experiments:

(i)   homogenisation mixing (HM; $\bar{G} = 300{-}500$ s^{-1}, mixing time 30–60 s),
(ii) rapid flocculation mixing (RM; $\bar{G} = 100{-}200$ s^{-1}, mixing time according to the operating parameters),
(iii) slow flocculation mixing (SM; $\bar{G} = 30{-}60$ s^{-1} in the case of Al-based coagulants, $\bar{G} = 50{-}80$ s^{-1} in the case of Fe-based coagulants, mixing time according to the operating parameters).

Rapid flocculation mixing does not always have to be followed by slow flocculation mixing. For example, if the conditions of a single-stage separation of the suspension using direct filtration are simulated, only rapid flocculation mixing is applied for a time corresponding to the real operating conditions.

It should also be noted that when simulating the conditions used at a particular water treatment plant, jar tests must be performed using identical reagents applied in the same forms, in the same orders, and at the same time intervals as at the water treatment plant. It is also necessary to ensure that the jar tests are carried out at the temperature of the raw water entering the treatment plant, as a different temperature could significantly affect the course of the jar test. The temperature mainly affects the dissociation of coagulant molecules and the subsequent course of metal hydrolysis (Xiao *et al.*, 2008) and thus the mechanisms of the interaction of coagulating components. Laboratory jar tests are very useful in testing and determining the required doses of coagulants, doses of reagents for adjusting the reaction pH, doses of flocculation aids, and their evaluation. Additionally, jar tests are useable for monitoring the effect of mixing conditions on the character of the resulting flocs when utilising mixing reactors with definable hydrodynamic conditions (e.g., a Taylor–Couette reactor). However, these laboratory tests are of limited use in cases where it is necessary to keep the similarity of model equipment with real scale operation, for example, when it is needed to precisely model the conditions of floc formation with respect to their separation by a specific technological process (sedimentation, flotation, filtration). An example situation is determining the effect of mixing conditions on the character of the resulting flocs and its impact on filtration cycles. In such cases, it is necessary to perform model measurements using a continuous-flow pilot plant that meets the required criteria of technological similarity with the actual real scale operation. Via pilot plant experiments, it is possible to accurately simulate the conditions of homogenisation, rapid mixing and slow flocculation mixing as well as single-stage and double-stage separation. Therefore, points 2 and 3 from the above optimisation scheme are not elaborated here in detail.

3.2 LABORATORY EQUIPMENT FOR JAR TESTS

The following equipment and material/chemicals are needed for conducting the jar tests:

(i)   Multi-position paddle stirrer – jar tester
      A paddle stirrer (jar tester) (Figure 3.1) must be adapted for positioning 1–2 L reaction vessels and equipped with suitable stirrers in each vessel position (see Section 3.3), a speed counter, an adjustable timer, and a speed control enabling a change in the stirring intensity in the range of the velocity gradient of approximately

Figure 3.1 A paddle stirrer (jar tester) LMK 8 (Institute of Hydrodynamics of the Czech Academy of Sciences) with eight 2 L jars.

$\bar{G} = 10$–500 s^{-1} (depending on the stirrers used). Four- or six-position paddle stirrers are commonly used in practice, but it is more advantageous to use eight- or ten-position paddle stirrers.

(ii) Reaction vessels (beakers, jars)

Reaction vessels should be of a circular cross-section (not a square cross-section) to ensure a steady distribution of the velocity gradient. All reaction vessels must have the same diameter, shape, and volume and be free of internal irregularities. Two litre vessels with a diameter of approximately 10 cm and a corresponding height of approximately 30 cm are optimal. Positions for individual vessels should be marked on the base of the jar tester to assure that the stirrers are located in their centres.

(iii) Reagent solutions

These are reagents for pretreatment of the coagulation pH, coagulants, or flocculation aids. Lime water (a saturated $Ca(OH)_2$ solution) or sodium hydroxide (NaOH) are usually used to adjust the pH to the alkaline range, and sulphuric acid (H_2SO_4) or hydrochloric acid (HCl) are usually used for acidification. Aluminium sulphate ($Al_2(SO_4)_3$) and ferric chloride or sulphate ($FeCl_3$ or $Fe_2(SO_4)_3$) are the most commonly used hydrolysing coagulants. Sometimes partially pre-polymerised coagulants such as PACl or PFS are applied. Synthetic polymers based on polyacrylamide or polyethyleneimine are mainly used as flocculation aids. All reagent solutions must be freshly prepared before conducting jar tests from the concentrated chemicals used at the corresponding treatment plant. The concentrations of the reagent solutions to be dosed during the laboratory jar tests should guarantee their stability and at the same time allow the highest possible dosing accuracy. The use of a 1% solution is recommended (mass fraction of solution = 0.01) for hydrolysing coagulants. The procedure for the preparation and dilution of coagulant solutions is given in Appendixes 4 and 5.

(iv) Additional laboratory equipment

Centrifugation may be applied for simulating the separation of aggregates via deep-bed (sand) filtration (Bubakova *et al.*, 2011; Naceradska *et al.*, 2019b; Pivokonsky *et al.*, 2009b). A laboratory centrifuge with a swinging rotor, a turning radius of 90 mm, and an adjustable speed of up to at least 4000 rpm is required. A suitable volume of centrifuge cuvettes is 150 mL or more. Other necessary equipment include pipettes, beakers, volumetric flasks, a pH meter, analytical instruments for monitoring the water quality parameters of interest, and so on.

3.3 STIRRERS FOR JAR TESTS AND INSTRUCTIONS FOR CALCULATING MIXING INTENSITY

Rotary mechanical stirrers are the most commonly applied stirrers for jar tests. There are two basic types: (i) a 'blade stirrer' with a single horizontal blade (Figure 3.2(a)) and (ii) a 'frame stirrer' with two vertical blades (Figure 3.2(b)). Cylindrical jars without baffles are usually used as reaction vessels with these stirrers (Polasek & Mutl, 1995).

The mixing intensity, characterised by the value of the mean velocity gradient $\bar{G}$, can be determined (provided that the power input of a stirrer P is known) as follows:

$$\bar{G} = \sqrt{\frac{P}{V\eta}}, \tag{3.1}$$

where P is the power input of the stirrer, V is the volume of the solution (water), and η is the dynamic viscosity of the solution (Appendix 1).

There are several ways to determine the power input of the stirrer. The most accurate is to determine the dependence of the torque M of a particular stirrer on its frequency. The power input of the stirrer P is calculated from the relationship

$$P = \omega M = 2\pi fM, \tag{3.2}$$

where ω is the angular velocity, M is the torque, and f is the stirrer frequency.

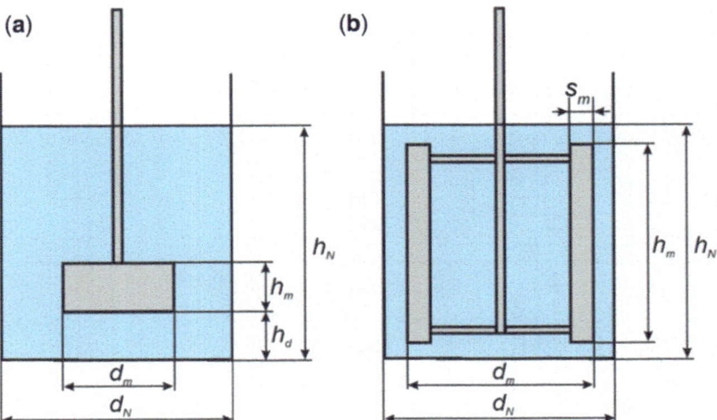

Figure 3.2 Rotary mechanical stirrers: (a) a blade stirrer with a single horizontal blade, (b) a frame stirrer with two vertical blades.

If a device is not available to measure the torque, the power input of the stirrer must be calculated from the relationship for the input power number Po

$$Po = \frac{P}{d_m^5 f^3 \rho_f}, \tag{3.3}$$

where P is the power input on the stirrer shaft, d_m is the diameter of the stirrer, f is the stirrer frequency, and ρ_f is the liquid density.

The characteristics for determining the power number Po of blade stirrers and frame stirrers are given in Table 3.1. A universal correlation equation can be used to determine the power number of these stirrers

$$Po = \left[\left(\frac{A_1}{Re} \right)^{A_2} + \frac{A_3}{Re^{\left(A_4 Re^{A_5} \right)}} + A_6 \right]^{1/A_2} \quad \text{for } Re < 2 \times 10^4, \tag{3.4}$$

where A_1–A_6 are empirically determined constants related to the specific geometry and dimensions of the stirrer given in Table 3.1, and Re is the Reynolds number, where $Re = ((\rho_f v d_m)/\eta)$ (in which ρ_f is the liquid density, v is the liquid flow rate, d_m is the diameter of the stirrer, and η is the dynamic viscosity of the liquid).

For the evaluation and/or optimisation of the mixing conditions (velocity gradient and mixing time) with regard to resultant floc properties, utilising a Taylor–Couette reactor might be suitable. This device consists of two concentric cylinders with one cylinder (usually the inner cylinder) rotating. An advantage of this arrangement is that it enables a relatively homogenous distribution of the velocity gradient within the liquid volume. Additionally, flocs formed in the Taylor–Couette reactor may be subsequently evaluated by non-destructive image analysis if required (Filipenska et al., 2019).

For the Taylor–Couette reactor, the velocity gradient value may be calculated as follows:

$$\bar{G} = \frac{2\omega R_1 R_2}{R_2^2 - R_1^2}, \tag{3.5}$$

where ω is the angular velocity, R_1 is the diameter of an inner cylinder, and R_2 is the diameter of an outer cylinder.

The flow regime between the two concentric cylinders in the Taylor–Couette reactor is characterised by the Reynolds number

$$Re = \frac{U \cdot (R_2 - R_1)}{\nu}, \tag{3.6}$$

where $R_2 - R_1$ is the width of the aperture between the two concentric cylinders, ν is the kinematic viscosity, and U is the mean flow velocity, defined as

$$U = \omega \frac{R_1 - R_2}{2}. \tag{3.7}$$

In the case of a static outer cylinder and a Newtonian fluid, increasing the inner cylinder angular velocity results in a gradual change of the flow regime from laminar Couette flow to turbulent flow (Kataoka, 1986).

Table 3.1 Characteristics of selected stirrers suitable for jar tests.

Stirrer	Diagram	Geometric parameters	Constants
Blade stirrer		$d_N/d_m = 2$ $h_m/d_m = 1$	$A_1 = 115$ $A_2 = 1.619$ $A_3 = 91.88$ $A_4 = 0.457$ $A_5 = 0$ $A_6 = 0$
Frame stirrer		$d_N/d_m = 1.11$ $h_m/d_m = 0.8$ $s_m/d_m = 0.12$ $h_d/d_m = 0.055$	$A_1 = 180$ $A_2 = 1.463$ $A_3 = 20.80$ $A_4 = 0.438$ $A_5 = 0$ $A_6 = 0$

3.4 WORKING PROCEDURE OF JAR TESTS

As already mentioned, a jar test can be performed in several modifications depending on the parameters to be optimised. The dose of a coagulant and the coagulation pH value are the fundamental parameters influencing the coagulation–flocculation of undesirable impurities in water and thus the overall efficiency of water treatment. In practice, a frequently applied procedure is to first determine the optimal dose of a coagulant by adding graduated amounts of this agent without pH control. Subsequently, the optimal pH value is determined by coagulation at different pH values within a certain range while using the determined optimal dose of the coagulant. This approach may lead to acceptable results, but it has one major drawback. Since hydrolysing coagulants (and to a lesser extent also pre-polymerised coagulants) lower the pH and $ANC_{4.5}$ of water depending on their dosed amount, it is not clear when conducting coagulant dose optimisation without pH control whether the best efficiency at the determined 'optimal' dose was truly due to the application of a suitable coagulant dose or because a suitable coagulation pH was reached at this dose (Naceradska *et al.*, 2019b). Therefore, simultaneous testing of both the coagulant dose and the coagulation pH appears to be a more accurate procedure. This must be performed in two steps. First, the effect of a coagulant dose on the change in the treated water pH, and thus the amount of a pH adjusting reagent (lime water/hydroxide/acid, see Section 3.2) required for reaching the target pH value, is determined. Then, coagulation tests are performed for individual doses of a coagulant in the pH range examined. A diagram of this optimisation procedure is shown in Figure 3.3.

3.4.1 Optimisation of coagulant dose and coagulation pH value

As described above, the addition of a coagulant affects the pH value of treated water. To reach the target coagulation pH values during a jar test (i.e., the pH values after coagulant addition), it is necessary to predetermine the required amounts of pH adjusting reagents. For this reason, preparative experiments (A) should be conducted prior to the optimisation jar tests (B).

(A) Determination of the doses of pH adjusting reagents:
   (1) Add a selected dose of a coagulant to a known volume of raw water (e.g., 250–1000 mL), mix the sample thoroughly (e.g., by using a magnetic stirrer) and measure the pH at the same time.
   (2) After the pH stabilises (under continuous stirring), gradually add small amounts of acid or base (depending on the target pH value) to adjust the pH of the sample; carefully record the added volumes of an acid/base and the resulting pH. The recommended volume of an acid/base added at a single time is in the range of some tenths of a mL. Follow the interval of, for example, 0.50, 0.25, or 0.10 mL of an acid/base and record the pH after each addition, then always wait for the pH value to stabilise. Continue until the border pH values (the lowest target pH in the case of adding an acid, or the highest target pH in the case of adding a base) that should be tested are reached.
   (3) Repeat this procedure for each dose of a coagulant to be tested (whose range is based on previous experience, relevant studies, etc.).
   The determined amounts of an acid/base will then be dosed into the raw water prior to coagulant addition to achieve the desired coagulation pH values during the jar tests (B). If the volume of the sample used for the determination of the dose of pH adjusting reagent and that of the sample used for the jar test are not the same, do not forget to recalculate the amounts of an acid–base correspondingly.

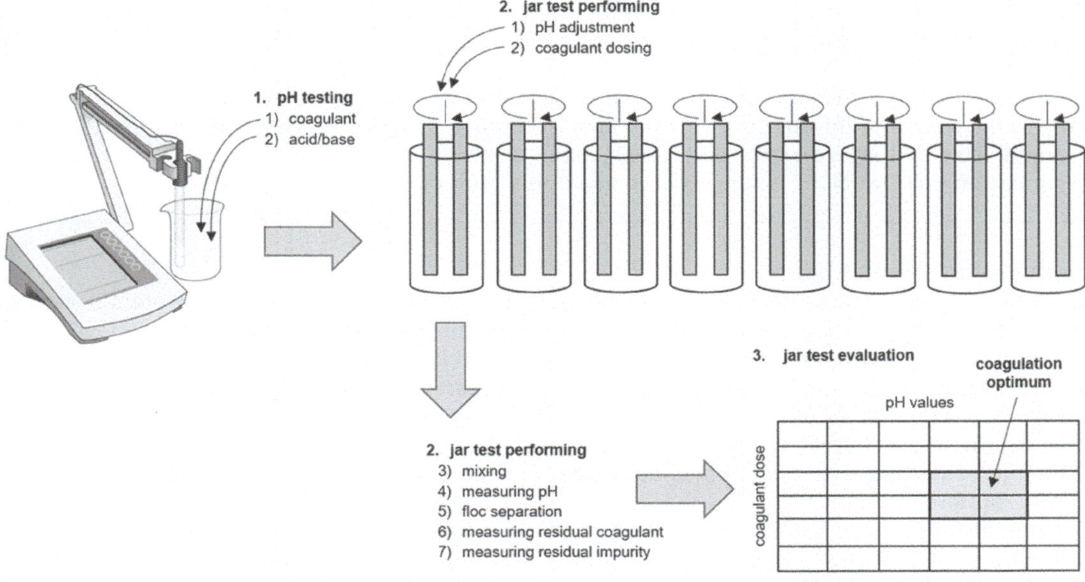

Figure 3.3 Schematic illustration of the jar test procedure and its evaluation (adapted from Naceradska *et al.*, 2019b).

SUPPLEMENTARY BOX

The addition of the most commonly used coagulants significantly affects the pH value of the water being treated. Therefore, the reported coagulation pH should always refer to the value after coagulant addition rather than the value before coagulation.

(B) Determination of the optimal coagulant dose and the optimal coagulation pH value via jar tests:
    (1) Fill numbered reaction vessels with raw water (up to approximately 3/4 of the required volume).
    (2) Add the pH adjusting reagents into the water in the reaction vessels according to the previously conducted experiments (A) so that there will be a different final pH in each vessel, covering the selected pH range. Stir the reagents in the vessels briefly.
    (3) Add more raw water to the reaction vessels to achieve the desired volume (e.g., 1 or 2 L).
    (4) Add the same dose of a coagulant to all reaction vessels. The same amount of a coagulant should also be dosed into a control sample of water without contaminants (e.g., deionised water) to subsequently check the concentration of the coagulant dosed.
    (5) Start homogenisation mixing ($t_{HM} = 30$–60 s) immediately after dosing the coagulant followed by rapid flocculation mixing corresponding to the desired value of velocity gradient ($\bar{G}_{RM}$), mix for the required rapid flocculation mixing time (t_{RM}).
    (6) After the rapid flocculation mixing time has elapsed, adjust the stirring speed so that it corresponds to the desired velocity gradient ($\bar{G}_{SM}$) value for slow flocculation mixing and mix for the desired slow flocculation mixing time (t_{SM}).
    (7) After the total desired mixing time has elapsed, it is advantageous to take a small sample from each vessel to immediately measure the pH to determine the coagulation pH value most precisely. Otherwise, the pH can be measured during mixing if jar tester and vessel construction allow.
    (8) Perform the separation of flocs by sedimentation and/or centrifugation (see Section 3.4.2).
    (9) After the separation of flocs is completed, take a water sample from each reaction vessel/centrifuge cuvette and determine the $ANC_{4.5}$, the residual concentration of a coagulant (usually the concentration of Fe or Al in case of hydrolysing and pre-polymerised coagulants), and the parameters characterising the target impurities, for example, turbidity in the case of inorganic particles, DOC or COD_{Mn} in the case of organic matter (UV_{254} can also be used for humic substances), cell numbers or optical density in the case of algal/cyanobacterial cells, and so on.; or conduct any more specific analysis if desired (e.g., the determination of pesticides, cyanobacterial toxins, etc.).
    (10) Repeat steps 1–9 for each dose of a coagulant to be investigated.
    (11) Record all input and measured parameters in previously prepared tables (as described in Section 3.4.3).

SUPPLEMENTARY BOX

When expressing a dose of a coagulant (particularly a hydrolysing metal salt) as a mass concentration, attention should be given to whether it is a hydrate or not, for example, ferric sulphate – $Fe_2(SO_4)_3$ or $Fe_2(SO_4)_3 \cdot 9H_2O$; ferric chloride – $FeCl_3$ or $FeCl_3 \cdot 6H_2O$; aluminium sulphate – $Al_2(SO_4)_3$ or $Al_2(SO_4)_3 \cdot 18H_2O$, and so on. Another possibility is to express a dose of a coagulant as a corresponding metal concentration, for example, Fe or Al. Molar concentrations can also be presented instead of mass concentrations. Conversion tables are provided in Appendixes 2 and 3.

COD_{Mn} is often utilised for the quantification of organic matter in drinking water. However, it should be noted that owing to the nature of this method, it is not able to capture all types of organic matter, particularly those of algal origin. Thus, measurement of only COD_{Mn} may result in an underestimation of the organic content. The determination of TOC/DOC is suggested to avoid this problem.

3.4.2 Floc separation after jar tests

The separation of flocs after the jar tests and before measuring the above-mentioned parameters depends on what type of separation is applied at the corresponding water treatment plant, that is, whether one-stage separation by direct filtration or two-stage separation by sedimentation followed by filtration.

Values expressing the water quality corresponding to the separation of flocs by **direct filtration** are measured after **centrifugation**, which takes place immediately after the end of mixing.

Values of the parameters characterising the quality of treated water after **sedimentation** are measured in samples taken after 60 min of **settling** employed immediately after the end of mixing; naturally, a different settling time can be chosen if appropriate.

Values expressing the water quality corresponding to the separation of flocs by **sedimentation followed by filtration** are measured in samples taken after 60 min of **settling** and subsequent **centrifugation**. In the case of optimisation involving dosing of a flocculation aid, separation is always performed by sedimentation with subsequent centrifugation (Polasek & Mutl, 1995).

Centrifugation must be performed under well-defined conditions using a swing rotor centrifuge at a centrifugal speed of 3500 rpm (1996 × g) for 20 min. Higher speeds and longer centrifugation times above the stated values do not affect the residual concentrations of the evaluated indicators (Hereit *et al.*, 1980).

3.4.3 Sampling and data recording

As described in Section 3.4.1, the pH should be measured immediately after the termination of mixing (or during coagulation, if possible); the presence of flocs does not affect the measured pH value significantly. However, samples for measurement of any other water quality parameters are collected after floc separation. It is necessary to always measure $ANC_{4.5}$ (to observe the consumption of alkalinity by the process), residual coagulant (there are strict threshold values for, e.g., the Al and Fe concentrations in treated water), and, naturally, residual content of all organic and/or inorganic impurities of interest. As already mentioned, this may include the determination of, for example, turbidity, DOC, COD_{Mn}, UV_{254}, cell numbers, optical density, pesticides, cyanobacterial toxins, and so on.

It is useful to record the values of the measured parameters in a table. Figure 3.4 can be used as a template. It contains data from three jar test series (eight jars each) conducted at a constant coagulant dose (50 mg L^{-1} $Fe_2(SO_4)_3$)

Title	Optimisation of pH value for the dose of 50 mg L^{-1} $Fe_2(SO_4)_3$		
Date	3rd September 2021		
Sample type/raw water	natural reservoir water		
Coagulant type	$Fe_2(SO_4)_3$		
Coagulant dose	50 mg L^{-1} $Fe_2(SO_4)_3$; corresponding to 13.97 mg L^{-1} Fe = 0.125 mmol L^{-1} Fe (dose 10 mL of 1% $Fe_2(SO_4)_3$ solution into 2 L)		
Homogenisation mixing (intensity and time)	100 rpm, i.e. $\bar{G}$ = 186 s^{-1}; 1 min		
Flocculation mixing – rapid/slow (intensity and time)	40 rpm, i.e. $\bar{G}$ = 47 s^{-1}; 30 min		

Notes:
- volume of water per vessel: 2 L
- raw water: temperature T = 13°C; $ANC_{4.5}$ = 1.82 mmol L^{-1}

jar	addition of pH adjusting agent (mL per jar)		pH (-)	$ACN_{4.5}$ (mmol L^{-1})	residual coagulant as Fe (mg L^{-1})	COD_{Mn} (mg L^{-1})	DOC (mg L^{-1})	DOC removal efficiency (%)
raw water		–	7.32	1.82	–	4.94	6.43	–
1		3.00	3.78	0.00	1.25	2.02	2.86	55.6
2		2.85	3.95	0.00	0.76	2.01	2.96	54.0
3	acidification with 0.5 M H_2SO_4	2.70	4.17	0.00	0.39	1.83	2.70	58.0
4		2.60	4.33	0.00	0.28	1.82	2.64	59.0
5		2.50	4.66	0.03	0.15	1.91	2.73	57.5
6		2.40	4.86	0.08	0.16	2.00	2.81	56.4
7		2.30	5.13	0.21	0.11	2.09	2.80	56.5
8		2.10	5.40	0.44	0.12	2.02	3.02	53.0
9		1.80	5.69	0.62	0.20	2.27	3.29	48.9
10		1.50	5.91	0.88	0.26	2.64	3.52	45.4
11		1.00	6.15	1.06	0.30	2.63	3.88	39.7
12		0.50	6.34	1.28	0.30	2.80	3.90	39.3
13		0.00	6.59	1.34	0.30	2.99	4.34	32.6
14		5.00	6.66	1.38	0.31	3.04	4.44	31.0
15		15.00	6.87	1.52	0.28	3.11	4.52	29.8
16		21.00	7.05	1.63	0.28	3.21	4.57	28.9
17	alkalization with $Ca(OH)_2$	27.00	7.21	1.79	0.29	3.30	4.93	23.4
18		30.00	7.35	1.83	0.29	3.39	5.09	20.9
19		33.00	7.47	1.95	0.29	3.49	5.20	19.2
20		35.00	7.57	2.03	0.28	3.56	5.24	18.6
21		40.00	7.89	2.12	0.24	3.64	5.21	19.0
22		45.00	8.48	2.66	0.21	3.67	5.41	15.9
23		50.00	8.91	2.98	0.17	3.75	5.48	14.8
24		55.00	9.15	3.12	0.17	3.75	5.38	16.4

Figure 3.4 An example of a jar test data record – investigating the effect of the coagulation pH value at a given coagulant dose (50 mg L^{-1} $Fe_2(SO_4)_3$) on the residual Fe, COD_{Mn}, and DOC; a part of the coagulation optimisation procedure for natural reservoir water.

and variable pH with relatively short intervals between the investigated pH values. In this case, ferric sulphate was used as the coagulant, and the target impurity was NOM; therefore, Fe, DOC, and COD_{Mn} were measured in the treated water. The table contains not only the determined values but also columns for recording the consumption of chemicals for pH adjustment and calculated DOC removal efficiencies in per cent.

To obtain the complete optimisation data, it would be necessary to repeat the same jar test series as shown in Figure 3.4 for other coagulant doses, for example, for 10, 30, 70 and 90 mg L^{-1} Fe$_2$(SO$_4$)$_3$. After acquiring the whole dataset, data processing and jar test evaluation follow as described in Section 3.4.4.

3.4.4 Data processing and evaluation

It is advantageous for the clear interpretation and presentation of jar test results to display them in the form of graphs or matrices that illustrate the dependence of the residual impurity and coagulant on the coagulation pH for distinct doses of the coagulants. For example, the DOC, COD_{Mn}, and Fe concentration results shown in Figure 3.4 are depicted as a graph in Figure 3.5. Another example is Figure 3.6, which depicts the dependencies of residual DOC and Al on the coagulation pH value at different coagulant (aluminium sulphate) doses. Example matrices (presenting the data shown in Figure 3.6 in a different way) are presented in Figure 3.7. Another example of result presentation via graphs/matrices is provided in Figures 3.8 and 3.9. In general, such graphic outputs improve the interpretation of the jar test results and help to reveal and explain the **optimal coagulant dose** and the **optimal coagulation pH range**.

The combination of the coagulant dose and the coagulation pH at which the maximum removal of impurities (displayed by the lowest residual content of, e.g., turbidity, DOC, etc.) is accompanied by the lowest coagulant residuals (e.g., Al or Fe) is usually considered the optimum. The region around this dose/pH, where the coagulation efficiency does not differ significantly, is the interval of the optimum.

Within the optimum interval, a suitable **operating dose** must then be determined. From an economic point of view, it is advantageous to use the minimum effective dose. However, the operating dose of a coagulant should not be selected at the boundary of the optimum interval, where a small change in raw water quality may cause coagulation to not occur. It is usually suitable to select the operating dose near the middle or in the lower half of the optimum interval (Pivokonsky *et al.*, 2011; Polasek & Mutl, 1995).

Additionally, the **operating pH** value should be chosen very carefully with respect to not only the maximum removal of contaminants but also the coagulant. Determination of the optimal coagulation conditions is further explained using some specific examples.

Figures 3.6 and 3.7 show the results of coagulation optimisation for water containing humic substances. Residual Al concentrations below the limit of 0.2 mg L^{-1} (drinking water threshold according to Council Directive 98/83/EC) were reached at the coagulation pH range of 5.8–6.2 and using the coagulant doses of 2.0 mg L^{-1} Al and higher. The highest DOC removal was reached at higher coagulant doses, while the optimal pH range broadened with increased dose. Nevertheless, the optimum intervals (for both the coagulant dose and the coagulation pH value) with respect to DOC and Al did overlap. As a result, the operating values could be a pH of approximately or slightly below 6 and a coagulant dose corresponding to approximately 3.5 mg L^{-1} Al.

Figures 3.8 and 3.9 depict the results of coagulation optimisation for water containing AOM, specifically the COM of cyanobacterium *Microcystis aeruginosa*. The optimum with respect to Al residuals was a pH between 6.2 and 6.8 at a dose of 6.5 mg L^{-1} Al, and with respect to DOC removal, the pH was between approximately 6.0 and 6.5 at the same dose. Thus, the optimum intervals again did overlap, and the operating conditions could be a pH of 6.5 and a coagulant dose corresponding to 6.5 mg L^{-1} Al.

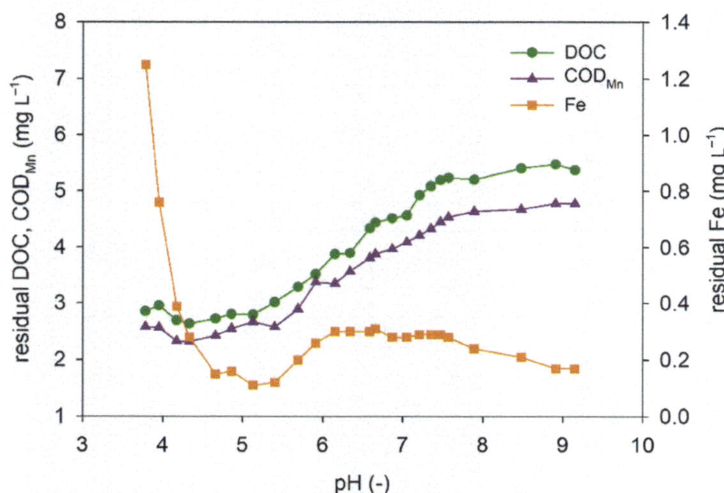

Figure 3.5 Dependence of the residual DOC, COD$_{Mn}$, and Fe on the coagulation pH value at a given coagulant dose (50 mg L^{-1} Fe$_2$(SO$_4$)$_3$); a part of the coagulation optimisation procedure for natural reservoir water.

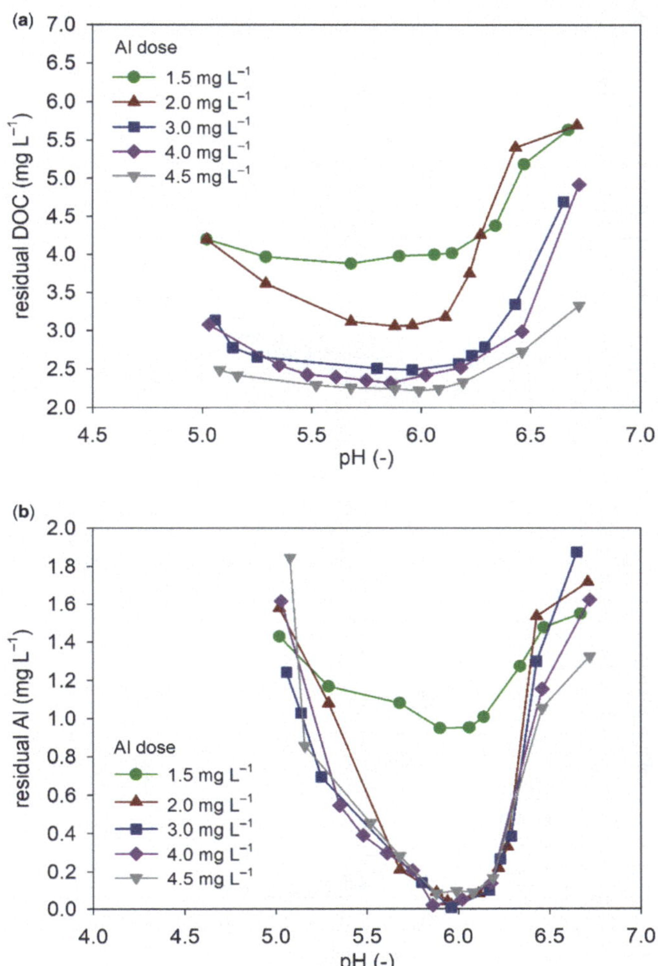

Figure 3.6 Dependence of (a) residual DOC and (b) residual Al on the coagulation pH value at different doses of the coagulant (aluminium sulphate at doses corresponding to 1.5–4.5 mg L⁻¹ Al); a part of the coagulation optimisation procedure for natural water with a high content of HS (initial DOC = 6.8 mg L⁻¹).

(a)

DOC mean removal efficiencies (%) – aluminium sulphate									
pH (-)									
5.0-5.3	5.3-5.6	5.6-5.8	5.8-5.9	5.9-6.1	6.1-6.2	6.2-6.3	6.3-6.4	6.4-6.5	6.5-6.7

Al dose (mg L⁻¹):

Al dose	5.0-5.3	5.3-5.6	5.6-5.8	5.8-5.9	5.9-6.1	6.1-6.2	6.2-6.3	6.3-6.4	6.4-6.5	6.5-6.7
1.5	38	42	43	41	41	41	38	36	24	17
2	38	47	54	55	55	53	45	35	20	16
3	54	61	62	63	63	62	61	59	51	31
4	55	62	65	66	64	63	61	59	56	28
4.5	63	65	67	67	67	66	65	62	60	51
%	70	67	65	60	55	50	40	30	20	15

(b)

Mean residual Al concentration (mg L⁻¹) – aluminium sulphate									
pH (-)									
5.0-5.3	5.3-5.6	5.6-5.8	5.8-5.9	5.9-6.1	6.1-6.2	6.2-6.3	6.3-6.4	6.4-6.5	6.5-6.7

Al dose (mg L⁻¹):

Al dose	5.0-5.3	5.3-5.6	5.6-5.8	5.8-5.9	5.9-6.1	6.1-6.2	6.2-6.3	6.3-6.4	6.4-6.5	6.5-6.7
1.5	1.43	1.17	1.08	0.95	0.95	1.01	1.13	1.27	1.48	1.55
2	1.58	1.08	0.21	0.09	0.02	0.08	0.22	0.33	1.54	1.72
3	1.24	0.69	0.14	0.10	0.01	0.10	0.27	0.39	1.30	1.87
4	1.62	0.55	0.30	0.02	0.05	0.14	0.31	0.75	1.15	1.62
4.5	1.85	0.86	0.28	0.08	0.10	0.17	0.29	0.69	1.05	1.32
	0.00	0.10	0.20	0.30	0.50	0.70	1.00	1.20	1.40	1.80

Figure 3.7 Dependence of (a) DOC removal efficiency and (b) residual Al on the coagulation pH value and the dose of the coagulant (aluminium sulphate at doses corresponding to 1.5–4.5 mg L⁻¹ Al); a part of the coagulation optimisation procedure for natural water with a high content of HS (initial DOC = 6.8 mg L⁻¹).

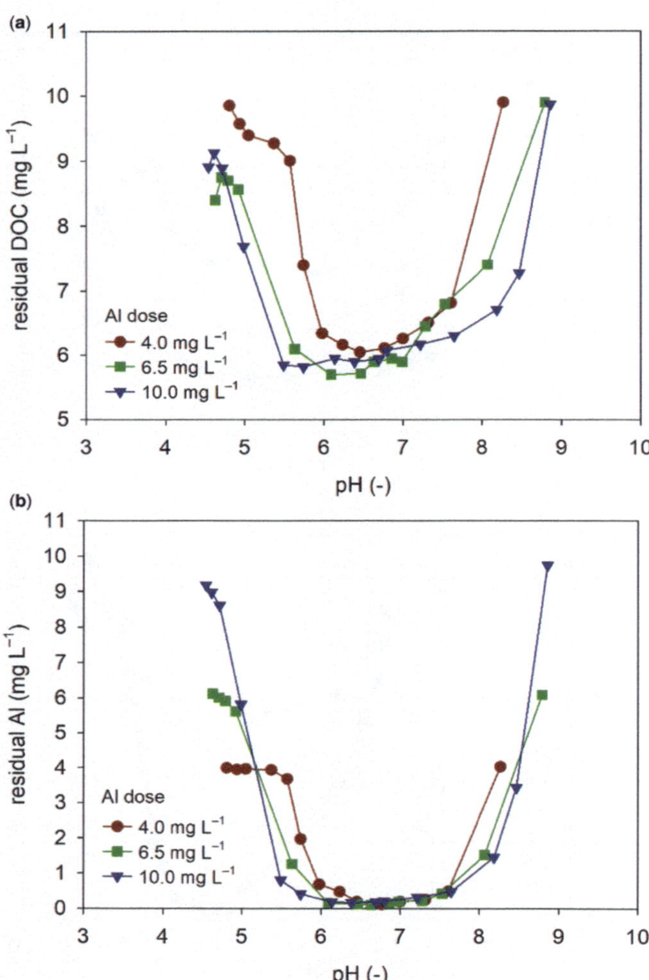

Figure 3.8 Dependence of (a) residual DOC and (b) residual Al on the coagulation pH value at different doses of the coagulant (aluminium sulphate at doses corresponding to 4–10 mg L^{-1} Al); a part of the coagulation optimisation procedure for water containing the COM of the cyanobacterium *Microcystis aeruginosa* (initial DOC = 10 mg L^{-1}).

(a) DOC mean removal efficiencies (%) – aluminium sulphate

Al dose (mg L^{-1})	pH (-)													
	4.8	4.9	5.1	5.4	5.6	5.7	6.0	6.3	6.5	6.8	7.0	7.3	7.6	8.3
4	1	4	6	7	10	26	37	38	40	39	37	35	32	1
6.5	13	14	19	30	39	41	43	43	43	41	41	36	31	12
10.5	11	21	25	36	42	42	42	41	41	39	39	38	37	29

%		45	43	41	39	37	31	19	14	6	1			

(b) Mean residual Al concentration (mg L^{-1}) – aluminium sulphate

Al dose (mg L^{-1})	pH (-)													
	4.8	4.9	5.1	5.4	5.6	5.7	6.0	6.3	6.5	6.8	7.0	7.3	7.6	8.3
4	4.00	3.95	3.97	3.95	3.69	1.98	0.69	0.48	0.20	0.12	0.20	0.26	0.50	4.04
6.5	5.92	5.50	4.82	2.86	1.63	1.27	0.16	0.15	0.13	0.13	0.20	0.23	0.43	2.82
10.5	8.61	6.35	5.82	2.02	0.69	0.41	0.28	0.18	0.17	0.19	0.25	0.36	0.48	2.02

0.00	0.10	0.20	0.30	0.40	0.70	1.00	1.60	2.00	3.00	4.00	4.80	5.50	6.50	8.60

Figure 3.9 Dependence of (a) DOC removal efficiency and (b) residual Al on the coagulation pH value and the dose of the coagulant (aluminium sulphate at doses corresponding to 4–10 mg L^{-1} Al); a part of the coagulation optimisation procedure for water containing the COM of the cyanobacterium *Microcystis aeruginosa* (initial DOC = 10 mg L^{-1}).

However, a more complicated situation occurs in the case that the coagulation conditions suitable for a pollutant removal do not overlap with the conditions that result in acceptable coagulant residuals. An example is shown in Figure 3.10(a). It depicts the coagulation optimisation procedure results for a natural reservoir water when using aluminium sulphate, the coagulant utilised at the DWTP supplied by the reservoir. It is apparent that the coagulation pH range suitable for DOC removal (~pH 5–6) was lower than the pH values at which satisfactory Al residuals were reached (around a pH of 7). Similar results were obtained at different aluminium sulphate doses. The DWTP was therefore forced to coagulate at pH 7 to meet the threshold values for residual Al, but the efficiency of DOC removal was only 20–30% at that pH value. In contrast, almost 50% of DOC was removed at a pH of 5.7. As shown in Figure 3.10(b), in this case, utilisation of ferric sulphate may be a solution for the DWTP. When using this coagulant, the optimal coagulation pH for DOC removal was approximately within the range of 4–5, while the lowest Fe residuals were reached around a pH of 5. Thus, when using ferric sulphate as the coagulant, an operational coagulation pH of 5 could be selected, where the residual Fe was below the threshold value and the DOC removal efficiency reached almost 60%.

For the raw water from the reservoir, another series of optimisation jar tests were conducted using both aluminium sulphate and ferric sulphate during different seasons of a year to capture the potential effect of possible changes in water quality. The results (shown as matrices) are depicted in Figures 3.11 and 3.12. Similar to the previous experiments, it was apparent that when using aluminium sulphate as the coagulant, the pH conditions at which the DOC residuals were the lowest (Figure 3.11(a)) did not overlap with the pH optimum based on the minimum Al residuals (Figure 3.11(b)). In contrast, when using ferric sulphate as the coagulant, the optimal coagulation pH with

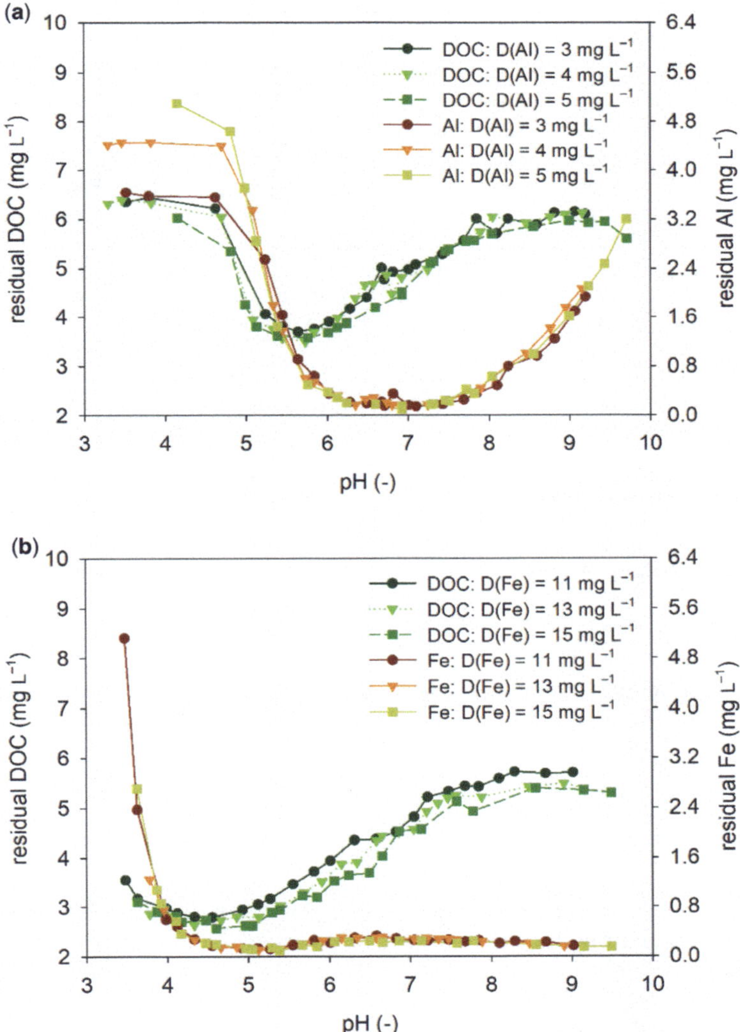

Figure 3.10 Dependence of (a) residual DOC and Al on the coagulation pH value at different doses of aluminium sulphate as the coagulant (corresponding to 3–5 mg L^{-1} Al) and (b) residual DOC and Fe on the coagulation pH value at different doses of ferric sulphate as the coagulant (corresponding to 11–15 mg L^{-1} Fe); a part of the coagulation optimisation procedure for natural water supplying a DWTP.

(a) Residual DOC concentration (mg L^{-1}) – aluminium sulphate

$Al_2(SO_4)_3$ dose (mg L^{-1})	pH (-)											
	4.75	5.00	5.25	5.50	5.75	6.00	6.25	6.50	6.75	7.00	7.50	8.00
10	8.02	7.20	5.86	5.36	5.03	5.01	5.31	5.73	6.06	6.65	7.16	7.79
15	8.07	6.31	5.50	5.52	5.55	5.01	5.19	5.79	5.36	5.61	7.21	8.59
20	7.21	6.27	5.21	5.23	5.01	5.02	5.37	5.78	6.40	6.31	7.09	7.41
25	7.58	6.71	5.18	4.70	4.67	4.82	4.67	5.33	6.05	6.55	7.01	7.00
30	6.97	5.78	4.66	4.66	4.68	4.66	4.85	5.07	5.65	5.90	6.30	6.53
35	6.87	5.34	4.59	4.51	4.54	4.66	4.69	4.96	5.71	6.18	6.50	6.45
40	6.22	5.06	4.67	4.60	4.71	4.81	5.06	5.19	5.25	5.67	6.17	6.90
45	5.59	5.09	4.57	4.59	4.63	4.67	4.74	5.29	5.28	5.33	6.43	6.53
50	5.65	4.50	4.19	4.25	4.45	4.60	4.70	4.89	5.40	5.73	5.74	6.68
60	5.59	4.58	4.37	4.32	4.50	4.43	4.44	4.76	4.99	5.08	6.02	6.40
70	4.89	4.58	4.23	4.38	4.33	4.55	4.50	4.72	4.73	5.19	6.03	6.47
80	5.07	4.51	4.15	4.30	4.24	4.24	4.44	4.55	4.67	5.94	5.99	6.36
90	4.54	4.47	4.25	4.21	4.30	4.46	4.43	4.52	4.63	4.96	5.78	6.07
100	4.27	4.31	4.11	4.01	4.20	4.27	4.52	4.77	4.76	4.94	5.65	6.13
%	3.00	4.00	4.50	5.00	5.50	6.00	6.50	7.00	7.50	8.00	8.50	9.00

(b) Residual Al concentration (mg L^{-1}) – aluminium sulphate

$Al_2(SO_4)_3$ dose (mg L^{-1})	pH (-)											
	4.75	5.00	5.25	5.50	5.75	6.00	6.25	6.50	6.75	7.00	7.50	8.00
10	1.765	1.686	1.733	1.284	0.565	0.357	0.133	0.107	0.105	0.134	0.371	0.674
15	1.776	2.763	1.774	1.123	0.408	0.167	0.097	0.085	0.095	0.099	0.438	1.474
20	3.127	2.424	1.372	0.910	0.376	0.250	0.129	0.110	0.124	0.144	0.342	0.562
25	4.514	2.905	1.879	1.084	0.590	0.217	0.158	0.094	0.137	0.243	0.345	0.634
30	4.764	4.009	1.288	0.745	0.567	0.207	0.098	0.089	0.092	0.130	0.211	0.456
35	5.693	2.173	1.702	0.878	0.313	0.147	0.106	0.129	0.095	0.168	0.302	0.854
40	6.075	4.623	2.579	0.725	0.247	0.114	0.131	0.088	0.085	0.105	0.229	0.920
45	5.857	3.478	1.483	0.834	0.264	0.104	0.151	0.082	0.083	0.113	0.369	0.621
50	6.770	2.675	1.239	0.578	0.273	0.278	0.144	0.066	0.096	0.111	0.376	0.882
60	7.194	4.244	2.185	1.202	0.570	0.327	0.102	0.053	0.071	0.087	0.278	0.468
70	4.412	3.322	1.546	0.841	0.396	0.242	0.105	0.065	0.056	0.085	0.253	1.003
80	5.963	3.189	1.256	0.731	0.444	0.108	0.103	0.065	0.071	0.065	0.569	1.648
90	6.075	3.862	1.698	0.877	0.572	0.189	0.118	0.053	0.052	0.066	0.318	1.422
100	4.447	3.447	1.845	0.549	0.316	0.183	0.124	0.133	0.056	0.151	0.256	0.838
%	0.000	0.050	0.100	0.150	0.200	0.300	0.500	1.000	3.000	5.000	7.000	8.000

Figure 3.11 Dependence of (a) DOC removal efficiency and (b) residual Al on the coagulation pH value and the dose of the coagulant ($Al_2(SO_4)_3$ at 10–100 mg L^{-1}, corresponding to 1.6–15.8 mg L^{-1} Al); a part of the coagulation optimisation procedure for natural water supplying a DWTP.

regard to both DOC removal (Figure 3.12(a)) and coagulant (Fe) residuals (Figure 3.12(b)) did overlap, as well as the suitable coagulant doses. Thus, the suggestion that the switch from aluminium sulphate to ferric sulphate would be beneficial was verified.

SUMMARISING BOX

The optimisation of a coagulant dose should not be conducted without pH control, as both of these parameters fundamentally affect the coagulation–flocculation efficiency. Since the addition of most coagulants significantly affects the pH, coagulation pH should always refer to the value after coagulation/after coagulant addition. The determination of the optimal coagulation conditions should be based on both the removal of the target impurities and on the coagulant residuals.

(a)

Fe$_2$(SO$_4$)$_3$ dose (mg L^{-1})	Residual DOC concentration (mg L^{-1}) – ferric sulphate										
	pH (-)										
	4.50	4.75	5.00	5.25	5.50	5.75	6.00	6.25	6.50	6.75	7.00
10	5.96	5.68	5.78	5.80	6.56	6.84	6.86	6.88	6.84	7.25	7.42
15	5.06	5.26	5.22	5.46	5.97	5.58	6.54	6.85	6.99	6.77	7.14
20	4.31	4.37	4.42	4.88	4.76	5.35	6.21	6.86	6.81	6.57	7.21
25	3.82	4.39	3.96	4.13	4.46	4.73	4.88	5.13	6.36	6.99	7.02
30	4.06	3.74	3.37	3.70	3.88	4.21	4.37	4.73	5.76	6.54	6.61
35	3.94	3.83	3.81	3.87	4.11	4.25	4.58	5.06	5.72	6.13	6.35
40	3.25	3.95	4.08	3.95	3.82	4.27	4.57	5.05	5.38	5.98	6.03
45	3.98	3.27	4.14	4.26	4.30	4.35	4.80	5.23	5.26	5.69	5.81
50	3.30	3.33	3.34	3.39	3.61	3.79	4.51	4.49	5.60	5.35	5.76
60	3.19	3.39	3.27	3.26	3.28	3.38	4.11	4.56	5.20	5.40	5.73
70	3.62	3.28	3.30	3.79	3.55	4.02	4.03	4.54	4.83	4.91	5.28
80	3.37	3.26	3.54	3.34	3.57	3.64	4.02	3.88	4.60	4.57	4.99
90	3.30	3.33	3.31	3.61	3.87	3.34	3.76	4.14	4.41	4.58	5.21
100	3.17	2.97	3.31	3.08	3.05	3.22	3.66	3.99	4.04	4.22	4.58
%	2.50	3.00	3.50	4.00	4.50	5.00	5.50	6.00	6.50	7.00	7.50

(b)

Fe$_2$(SO$_4$)$_3$ dose (mg L^{-1})	Residual Fe concentration (mg L^{-1}) – ferric sulphate										
	pH (-)										
	4.50	4.75	5.00	5.25	5.50	5.75	6.00	6.25	6.50	6.75	7.00
10	1.097	0.646	0.862	0.715	0.872	1.400	1.494	1.373	0.856	0.754	0.507
15	0.420	0.422	0.420	0.469	0.654	0.998	1.078	1.021	0.874	0.663	0.619
20	0.481	0.222	0.229	0.263	0.296	0.448	0.853	1.164	0.984	0.577	0.763
25	0.282	0.243	0.190	0.189	0.188	0.250	0.325	0.408	0.779	0.520	0.490
30	0.284	0.260	0.175	0.173	0.164	0.205	0.337	0.465	0.584	0.516	0.418
35	0.242	0.189	0.167	0.158	0.143	0.186	0.273	0.460	0.576	0.471	0.468
40	0.327	0.145	0.117	0.119	0.115	0.165	0.276	0.367	0.372	0.479	0.454
45	0.249	0.167	0.140	0.128	0.127	0.171	0.279	0.362	0.469	0.518	0.392
50	0.188	0.111	0.126	0.126	0.118	0.152	0.291	0.325	0.386	0.358	0.371
60	0.264	0.142	0.146	0.106	0.105	0.121	0.192	0.246	0.446	0.476	0.487
70	0.162	0.143	0.118	0.136	0.143	0.154	0.188	0.237	0.400	0.473	0.499
80	0.189	0.154	0.139	0.127	0.146	0.164	0.193	0.216	0.310	0.476	0.455
90	0.156	0.129	0.133	0.136	0.142	0.149	0.183	0.261	0.332	0.278	0.440
100	0.196	0.129	0.151	0.125	0.133	0.129	0.129	0.178	0.180	0.207	0.193
%	0.000	0.050	0.100	0.150	0.200	0.500	0.800	1.000	1.200	1.400	1.500

Figure 3.12 Dependence of (a) DOC removal efficiency and (b) residual Fe on the coagulation pH value and the dose of the coagulant (Fe$_2$(SO$_4$)$_3$ at 10–100 mg L^{-1}, corresponding to 2.8–27.9 mg L^{-1} Fe); a part of the coagulation optimisation procedure for natural water supplying a DWTP.

3.5 TEST OF FLOCCULATION

The test of flocculation is a simple and quick procedure for evaluating the coagulation–flocculation effectiveness and the general quality of the formed suspension (aggregates–flocs) (Bubakova *et al.*, 2011; Hereit *et al.*, 1980). This procedure categorises the formed flocs into four groups: **macro-flocs**, **micro-flocs**, **primary flocs**, and **non-flocculated portion** on the basis of their sedimentation velocities. Flocs belonging to these groups differ significantly in their properties, and thus, they are suitable for different separation methods (see Table 3.2). A big advantage of this test is its simplicity. It can be conducted either in a laboratory after conducting jar tests or directly at a treatment plant so as to evaluate the flocculation efficiency within the distinct parts of the facility, for example, after mixing, within the sedimentation tank, after sand filtration, and so on. The required instrumentation consists only of a centrifuge and equipment for the analysis of the monitored components, that is, the target impurities and the utilised coagulant.

The individual particle categories are defined as follows.

Macro-flocs (MA) are flocs that are removed by simple sedimentation in less than 5 min. Their share (P$_{MA}$) is determined as the ratio of the difference between the concentration C_0 of the monitored parameter determined at

Table 3.2 Characteristics of the individual types of flocs, determination of their share in the suspension and suitable separation technologies.

Group Name	Macro-flocs	Micro-flocs	Primary Flocs	Non-flocculated Portion
Size of flocs	>1 mm	0.05–1 mm	0.005–0.05 mm	<0.005 mm
Other properties of flocs	Turbidity is evident, as in the case of micro-flocs, but morphological properties vary.	Turbidity appears. Orthokinetic motion predominates.	Opalescence appears. Orthokinetic motion predominates.	These particles do not affect a beam of light passing through the system. Perikinetic motion predominates.
Characteristics of flocs in terms of the flocculation test	Particles that settle in less than 5 min.	Particles that settle in between 5 and 60 min.	Particles that settle between 60 and 440 min. Long-lasting sedimentation in the gravity field was replaced by sedimentation in a centrifugal field.	Particles that are not affected by the gravitational field and do not settle even after a very long time.
Share of flocs – calculation[*]	$P_{MA} = \dfrac{C_0 - C_5}{C_0}$	$P_{MI} = \dfrac{C_5 - C_{60}}{C_0}$	$P_{PR} = \dfrac{C_{60} - C_{F(60)}}{C_0}$	$P_{NA} = \dfrac{C_{F(60)}}{C_0}$
Suitable separation technology	Sedimentation and filtration	Double-stage filtration (clarification and filtration)	Direct filtration	Coagulation (aggregation) filtration
Description of the technology and ongoing processes	Homogenisation of a coagulant, formation of micro- and macro-flocs during rapid and slow flocculation mixing, sedimentation of the formed flocs and separation of the remaining particles on a filter.	Homogenisation of a coagulant, simultaneous flocculation and filtration of micro-flocs in a perfectly floating flocculent cloud (usually in a clarifier) and separation of the remaining particles on a filter.	Homogenisation of a coagulant, formation of primary flocs in a rapidly mixed tank, inlet to a filter. Adhesion of destabilised particles to the surface of a filtration material.	Homogenisation of a coagulant and immediate flow of water to a filter. Adhesion of destabilised particles to the surface of a filtration material.

$*C_0$ is the total initial concentration of Al or Fe at the beginning of sedimentation, C_5 and C_{60} are the concentrations after 5 and 60 min of sedimentation, respectively, and $C_{F(60)}$ is the concentration after 60 min of sedimentation and subsequent centrifugation (3500 rpm $= 1996 \times g$, 20 min). Centrifugation conditions were determined by testing suspensions under different conditions (speed and time). The non-flocculated portionno longer changes with increasing the speed and lengthening the time above the stated values.

the beginning of sedimentation and concentration C_5 in the sample taken after 5 min of sedimentation $(C_0 - C_5)$ to the concentration C_0

$$P_{MA} = \frac{C_0 - C_5}{C_0}. \tag{3.8}$$

Micro-flocs (MI) are flocs that are removed by simple sedimentation between 5 and 60 min. Their share (P_{MI}) is determined as the ratio of the difference between the concentration of the monitored parameter determined in the samples after 5 and 60 min of sedimentation $(C_5 - C_{60})$ to the concentration C_0

$$P_{MI} = \frac{C_5 - C_{60}}{C_0}. \tag{3.9}$$

Primary flocs (PR) are flocs that are removable, for example, by direct filtration, but their sedimentation time is longer than 60 min. To accomplish the flocculation test within a reasonable time, their removal is simulated via centrifugation under defined conditions (3500 rpm (1996 × g) for 20 min). The share of primary flocs (P_{PR}) is then determined as the ratio of the difference between the concentration C_{60} of the monitored parameter after 60 min of sedimentation and the concentration $C_{F(60)}$ in the centrifuged sample after 60 min of sedimentation $(C_{60} - C_{F(60)})$ to the concentration C_0

$$P_{PR} = \frac{C_{60} - C_{F(60)}}{C_0}. \tag{3.10}$$

Non-flocculated portion (NF) corresponds to the fraction that is not removable via coagulation–flocculation, and its share (P_{NF}) is determined as the ratio of the concentration $(C_{F(60)})$ of the monitored parameter in the centrifuged sample after 60 min of settling to the concentration C_0

$$P_{NF} = \frac{C_{F(60)}}{C_0}. \tag{3.11}$$

Other important parameters describing the quality of the suspension formed during coagulation–flocculation are the **degree of destabilisation** and the **degree of aggregation**.

The **degree of destabilisation** (α_D) is defined as the ratio of the number of destabilised particles to the total number of particles. The value of α_D can range from 0 to 1; $\alpha_D = 0$ means that destabilisation did not occur, while $\alpha_D = 1$ means that all the particles present in the water were destabilised (this is rather a theoretical value not attainable under real conditions). In practice, the value of α_D is determined as the ratio of the difference between the initial concentration C_0 of a selected parameter and the concentration $C_{F(HM)}$ in a centrifuged sample collected after homogenisation mixing $(C_0 - C_{F(HM)})$ to the concentration C_0

$$\alpha_D = \frac{C_0 - C_{F(HM)}}{C_0}. \tag{3.12}$$

The **degree of aggregation** (α_A) is defined as the ratio of the number of aggregated particles to the total number of particles. Similar to α_D, the value of α_A can range from 0 to 1; $\alpha_A = 0$ means that aggregation did not occur, while $\alpha_A = 1$ means that all the particles present in the water were aggregated (which is again only a theoretical situation). In practice, α_A is determined as the ratio of the difference between the initial concentration C_0 of a selected parameter and the concentration $C_{F(A)}$ in a centrifuged sample collected after flocculation mixing $(C_0 - C_{F(A)})$ to the concentration C_0

$$\alpha_A = \frac{C_0 - C_{F(A)}}{C_0}. \tag{3.13}$$

There are many other more advanced methods applicable for detailed evaluation of the floc properties, including not only their size and/or sedimentation velocity but also their structure, shape, density, porosity, and so on. However, elaboration of those methods is beyond the scope of this publication, and the readers (if interested) are instead referred to the literature (e.g., Bagheri *et al.*, 2015; Liang *et al.*, 2015; Merkus, 2009). Nevertheless, it is generally beneficial to apply methods that can be performed *in situ* and do not require collecting and transportation of the sample, since these steps may alter the floc properties.

3.5.1 Flocculation test working procedure

One possibility is to perform the flocculation test following a laboratory jar test conducted under given conditions. Then, the procedure is as follows.

(1) Conduct a jar test under selected conditions that correspond to the purpose of the optimisation procedure – for example, apply the optimal coagulant dose and the optimal coagulation pH value and test different velocity gradients, or different times of slow flocculation mixing, and so on.
(2) Take a sample from each reaction vessel immediately at the end of mixing to determine the concentration of the basic component of the coagulant (Al/Fe) – concentration C_0.
(3) Let the flocs in the reaction vessels settle, do not move the reaction vessels.

(4) Take a sample from each reaction vessel after 5 min of settling to determine the concentration of the basic component of the coagulant (Al/Fe) – concentration C_5.

(5) Take a sample from each reaction vessel after 60 min of settling to determine the concentration of the basic component of the coagulant (Al/Fe) – concentration C_{60}.

(6) Take a sample from each reaction vessel immediately after the sampling described in point 5 (therefore after 60 min of sedimentation) for the subsequent centrifugation; then determine the concentration of the basic component of the coagulant (Al/Fe) in the centrifuged sample – concentration $C_{F(60)}$.

(7) Take a sample from each reaction vessel for the determination of other monitored parameters.

SUPPLEMENTARY BOX

Attention should be given to a proper quantification of mixing intensity. In the literature, it is often expressed as a rotation frequency of a certain impeller in revolutions per minute (rpm). However, rpm units provide no information about the hydrodynamics. It is thus always necessary to provide the value of the global velocity gradient ($\bar{G}$) instead of the rpm.

Title	Optimisation of velocity gradient for the dose of 15 mg L^{-1} $Al_2(SO_4)_3 \cdot 18\ H_2O$
Date	13th May 2010
Sample type/raw water	natural water – Flaje reservoir (Meziboří, Czech Republic)
Coagulant type	$Al_2(SO_4)_3 \cdot 18\ H_2O$
Coagulant dose	15 mg L^{-1} $Al_2(SO_4)_3 \cdot 18\ H_2O$; corresponding to 1.21 mg L^{-1} Al = 0.023 mmol L^{-1} Al
Flocculation mixing – rapid/slow (intensity and time)	$\bar{G}$ = 20–400 s^{-1}; t = 150 s

Notes:
- volume of water per vessel: 2 L
- raw water: temperature T = 8°C; $ANC_{4.5}$ = 0.18 mmol L^{-1}

jar	$\bar{G}$ (s^{-1})	C_0 (mg L^{-1})	C_5 (mg L^{-1})	C_{60} (mg L^{-1})	$C_{F(60)}$ (mg L^{-1})	$\alpha_{A(Al)}$ (-)
1	20	1.197	0.896	0.443	0.205	0.829
2	40	1.205	0.875	0.434	0.171	0.858
3	60	1.210	0.909	0.533	0.122	0.899
4	80	1.198	0.960	0.515	0.081	0.932
5	100	1.211	1.032	0.702	0.091	0.925
6	150	1.188	1.067	0.748	0.084	0.929
7	200	1.199	1.162	0.983	0.062	0.948
8	250	1.193	1.156	1.014	0.026	0.978
9	300	1.204	1.169	1.072	0.013	0.989
10	350	1.209	1.174	1.100	0.010	0.992
11	400	1.195	1.162	1.064	0.001	0.999

jar	$\bar{G}$ (s^{-1})	P_{MA} (-)	P_{MI} (-)	P_{PR} (-)	P_{NF} (-)	$\alpha_{A(Al)}$ (-)
1	20	0.171	0.199	0.379	0.251	0.829
2	40	0.142	0.218	0.366	0.274	0.858
3	60	0.101	0.339	0.311	0.249	0.899
4	80	0.068	0.362	0.371	0.199	0.932
5	100	0.075	0.505	0.272	0.148	0.925
6	150	0.071	0.559	0.268	0.102	0.929
7	200	0.052	0.768	0.149	0.031	0.948
8	250	0.022	0.828	0.119	0.031	0.978
9	300	0.011	0.879	0.081	0.029	0.989
10	350	0.008	0.902	0.061	0.029	0.992
11	400	0.001	0.889	0.082	0.028	0.999

Figure 3.13 An example flocculation test data record – investigating the effect of the velocity gradient at a fixed mixing time on the size distribution of the flocs. The Al concentrations determined in the collected samples are recorded in the first part; in the second part, the calculated shares of the different groups of flocs (P_{MA}, P_{MI}, P_{PR}, P_{NF}) and α_A values are included.

The samples taken for the determination of the Al/Fe concentrations after certain sedimentation times (C_0, C_5, and C_{60}) are always collected at a depth of 40 mm below the water surface directly from the reaction vessels with a pipette. Attention should be given not to disturb the sedimentation process while collecting the samples.

Centrifugation (for the determination of $C_{F(60)}$, point 6) is carried out under the same conditions as in the case of optimisation jar tests described above, that is, using a centrifuge with a swinging rotor operated at a centrifugal speed of 3500 rpm ($1996 \times g$) for 20 min.

The relative share of the listed groups of flocs is determined from equations (3.8–3.11). Table 3.2 provides detailed information on the properties of the flocs and appropriate separation technologies.

The procedure described above is related to laboratory coagulation–flocculation optimisation via jar tests. However, as already mentioned, flocculation test can also be conducted at a drinking water treatment plant. In such a case, suitable points of the treatment chain are selected, water samples from these points are collected in jars or beakers of sufficient volume (e.g., 2 L), and then the procedure continues according to points 2–7.

3.5.2 Data recording, processing and evaluation

An example of a flocculation test data record is shown in Figure 3.13. It includes the results of testing different velocity gradients at a fixed mixing time, while Al was the monitored parameter owing to the utilisation of aluminium sulphate as the coagulant. In the first part of the list of results, the Al concentrations determined in the collected samples are recorded; in the second part, the calculated shares of the different groups of flocs (P_{MA}, P_{MI}, P_{PR}, P_{NF}) and α_A values are presented. The same data are displayed visually as a part of Figure 3.14, which shows the floc proportion distributions for the different velocity gradients and different mixing times. Similar to the coagulant dose and coagulation pH optimisation, the graphical presentation of the flocculation test results contributes to their clearer interpretation and evaluation. It is obvious from Figure 3.14 that in this case, the

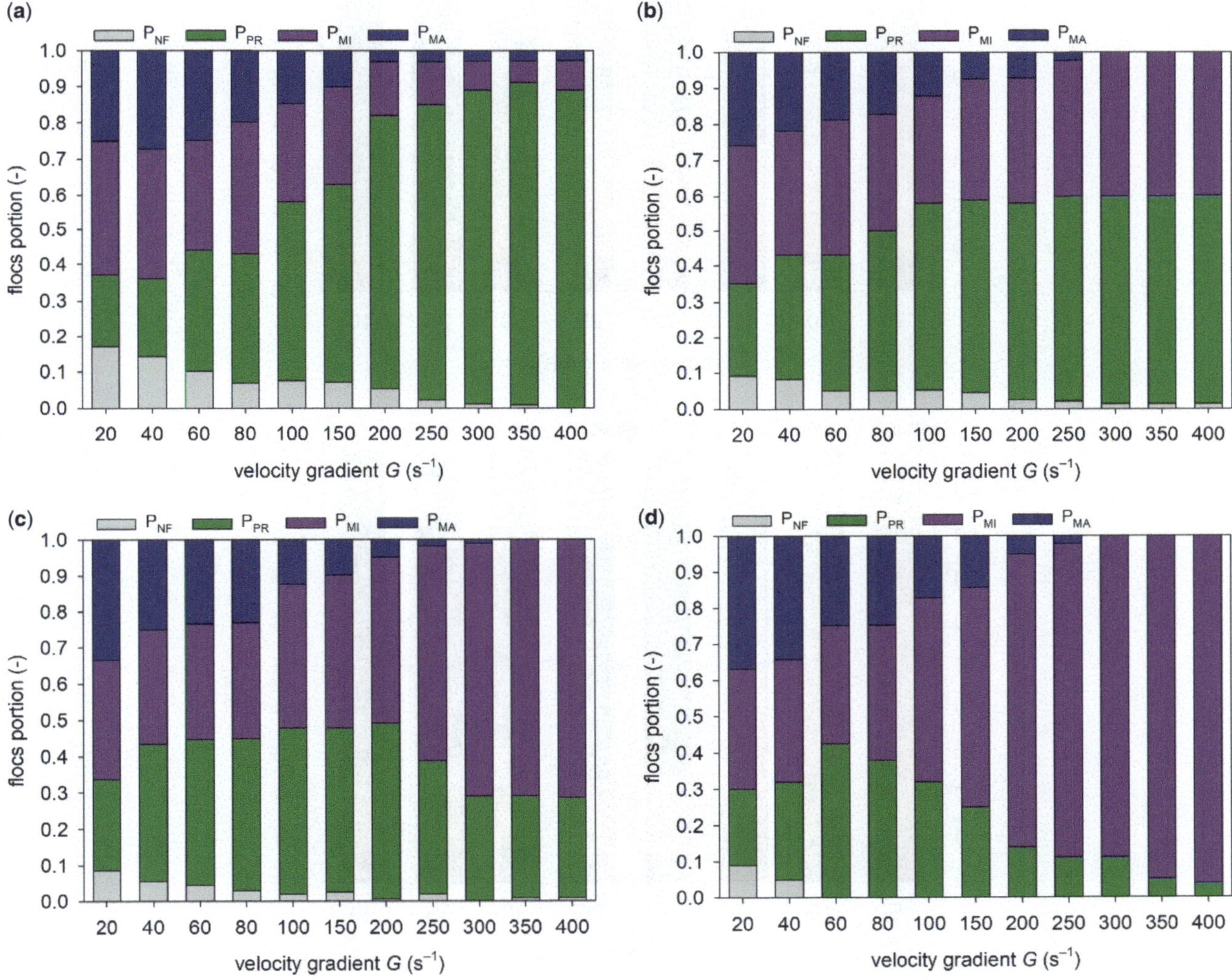

Figure 3.14 Relative portions of the different groups of flocs depending on the velocity gradients applied at mixing times of (a) 150 s, (b) 300 s, (c) 600 s, and (d) 900 s.

shares of the non-flocculated portion (P_{NF}) and the share of macro-flocs (P_{MA}) generally decreased with increasing $\bar{G}$, while the values also depended on the mixing time. Similarly, the shares of micro-flocs (P_{MI}) and primary flocs (P_{PR}) that together comprised the majority varied both with the $\bar{G}$ value and mixing time. Notably, the determination of the optimal mixing conditions with regard to the size distribution of arising flocs would always be dependent on the type of floc separation operated by a particular water treatment plant (i.e., sedimentation, direct sand filtration, etc.), and it is therefore not possible to select the optimum without referring to the broader context.

Figures 3.15 and 3.16 further illustrate the results of the laboratory flocculation optimisation procedure for a drinking water treatment plant that operates single-stage floc separation via direct sand filtration. Figure 3.15(a), (b) show the dependence of the floc proportion distribution on the mixing time when using two different coagulants (PAX-18 and aluminium sulphate) under optimised coagulation conditions. In the case of PAX-18, it is apparent that an application of a mixing time of 15 min or longer would be inappropriate due to the significant proportion of macro-flocs (over 50%) that are unsuitable for direct sand filtration. Similarly, in the case of aluminium sulphate, a mixing time of 20 min or longer would be too long for the same reason.

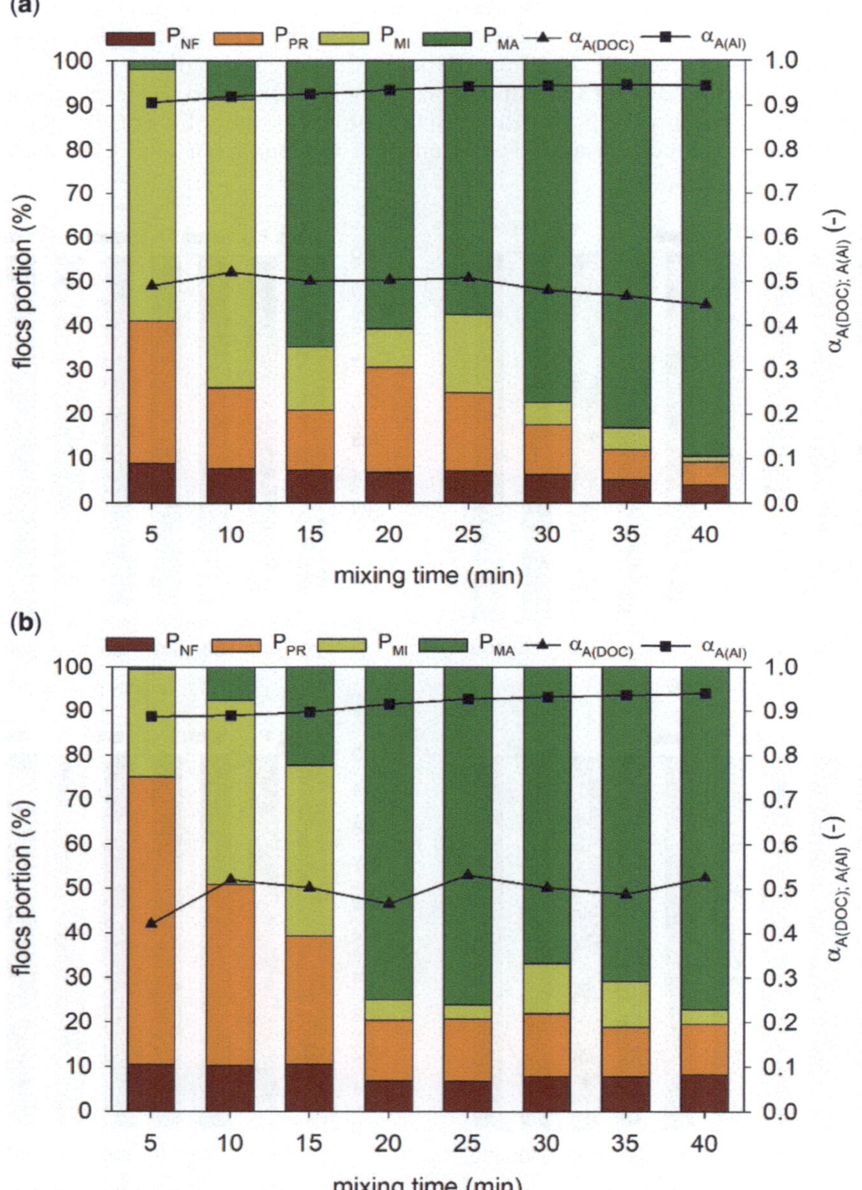

Figure 3.15 Relative portions of the different groups of flocs and the values of the coefficients of aggregation depending on the mixing time when using (a) PAX-18 and (b) aluminium sulphate as the coagulants under optimised coagulation conditions.

Figure 3.16(a), (b) illustrate the relationship between the floc proportion distributions and the velocity gradient at a given mixing time (10 min). When using PAX-18, the undesirable share of macro-flocs was low at $\bar{G}$ 20 s⁻¹, then sharply increased at $\bar{G}$ 30 s⁻¹, and then gradually decreased up to $\bar{G}$ 200 s⁻¹. A suitable floc size distribution was therefore reached at $\bar{G}$ 20 s⁻¹ and at $\bar{G}$ 200 s⁻¹ or higher. When using aluminium sulphate, the share of macro-flocs increased with $\bar{G}$ up to $\bar{G}$ 150 s⁻¹ and then sharply dropped; nevertheless, the share of macro-flocs was generally lower than when applying PAX-18. The values of $\alpha_{A(DOC)}$ and $\alpha_{A(Al)}$ were approximately similar for both coagulants. Thus, the utilisation of aluminium sulphate as a coagulant appeared to be more appropriate owing to its lower tendency to form undesirable macro-flocs at a suitable mixing time and over a wide range of velocity gradients. If PAX-18 was employed as a coagulant, particular attention would have to be paid to applying a suitable velocity gradient to avoid the formation of macro-flocs.

Figures 3.17 and 3.18 show the results of the flocculation tests conducted at an operating water treatment plant. Their aim was to assess the properties of flocs formed at given stages of the treatment chain (within a flocculation

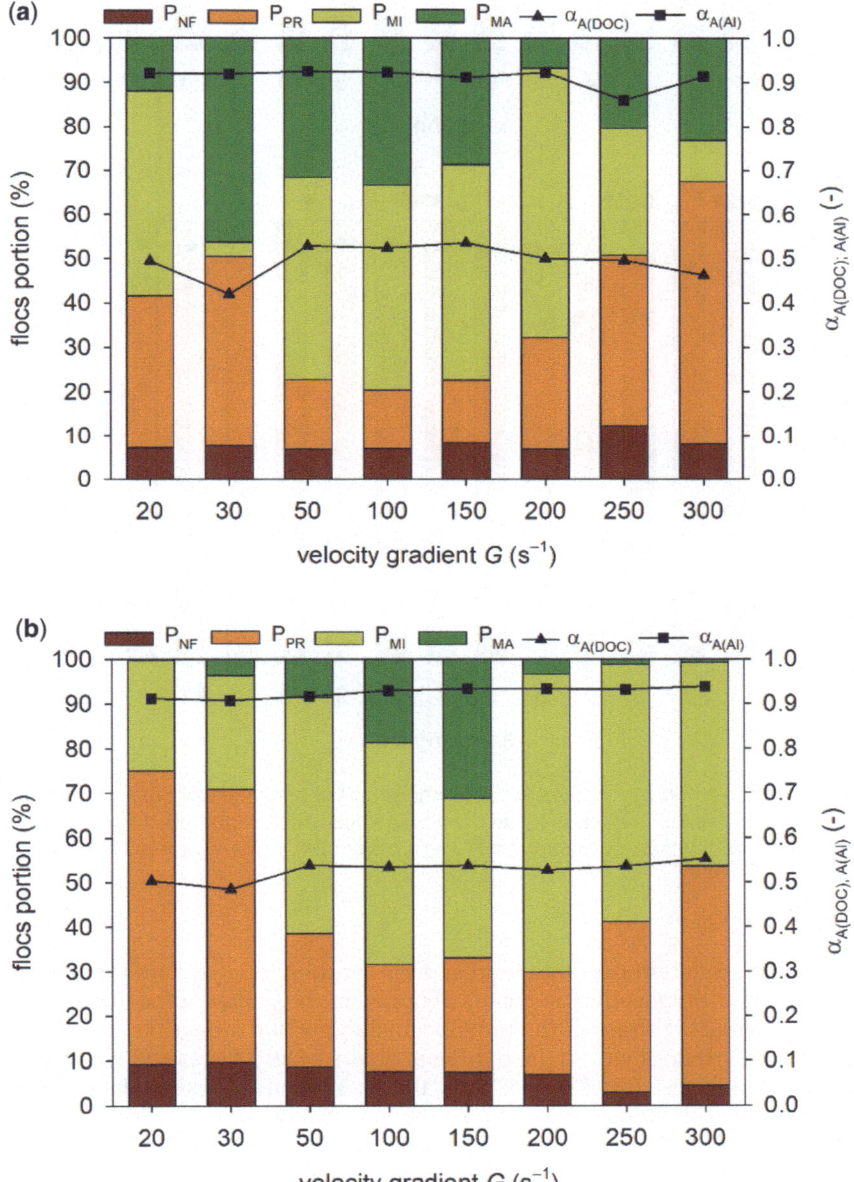

Figure 3.16 Relative portions of the different groups of flocs and the values of the coefficients of aggregation depending on the velocity gradient with a mixing time of 10 min when using (a) PAX-18 and (b) aluminium sulphate as the coagulants under optimised coagulation conditions.

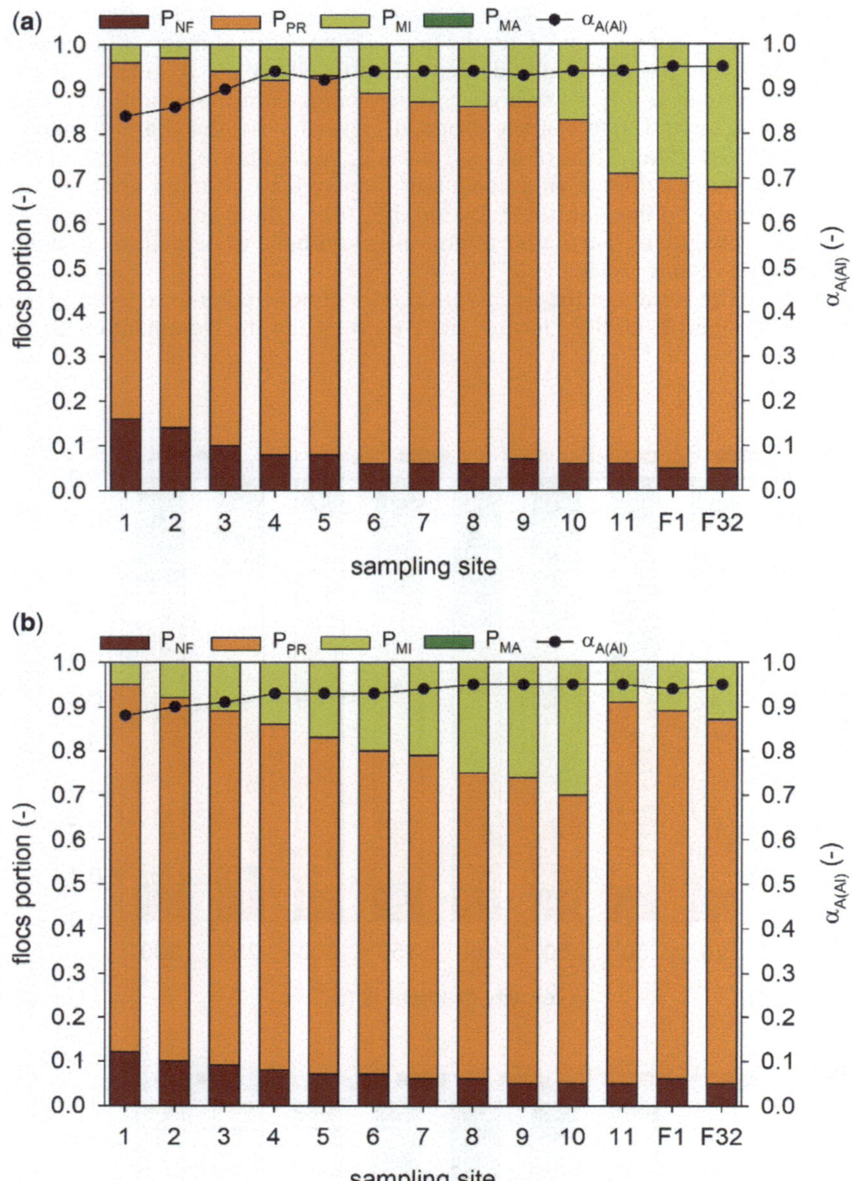

Figure 3.17 Relative portions of the different groups of flocs and the values of the coefficient of aggregation at given sampling sites throughout a DWTP treatment chain (sites 1–10 were located within a flocculation canal; site 11 was at the entrance to the filters; sites F1 and F32 were the selected filters). Flocculation tests were conducted in the winter at flow rates of (a) 1.5 m³ s⁻¹ and (b) 3.5 m³ s⁻¹.

canal and at selected filters) at different flow rates and at different times of a year (in the winter and spring because the raw water quality changes with season). The results obtained in the winter at the lower flow rate (Figure 3.17(a)) showed a clear trend of a decreasing share of the non-flocculated portion alongside increasing values of $\alpha_{A(Al)}$ and an increasing proportion of micro-flocs within the treatment chain. Macro-flocs were not formed, and primary flocs prevailed at all of the treatment stages investigated. Thus, this was found to be suitable for the direct sand filtration method of separation operated at the treatment plant. At the higher flow rate, the results were quite similar (Figure 3.17(b)), with the exception of partial floc breakage that occurred prior to entering the filters, which was indicated by a decrease in the P_{MI} and a relative increase in the P_{PR}. Nevertheless, in the winter, the primary flocs that were optimal for direct filtration clearly prevailed at both of the applied flow rates.

The results obtained in spring at the lower flow rate (Figure 3.18(a)) showed some similarities to the winter results, that is, the share of the non-flocculated portion generally decreased, the $\alpha_{A(Al)}$ value increased, and the proportion of micro-flocs increased throughout the treatment chain, while the primary flocs prevailed. However, the share of

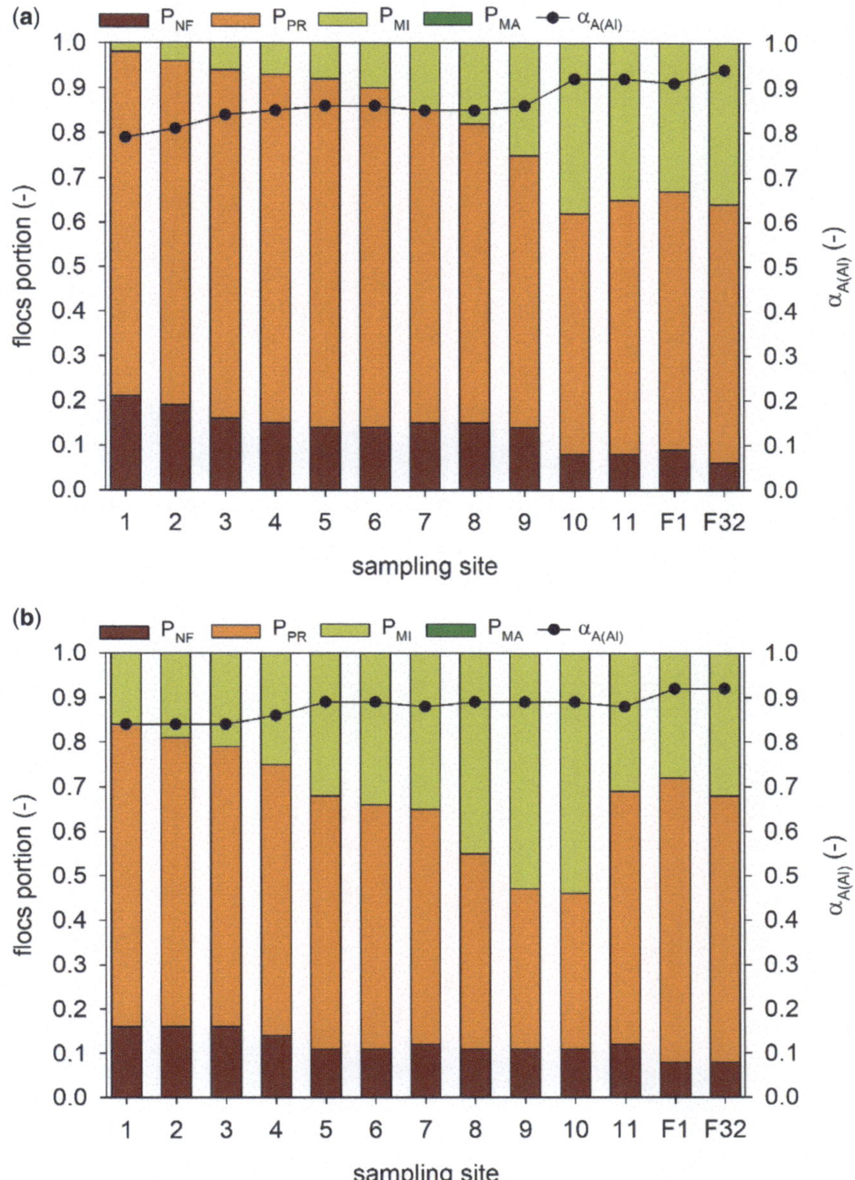

Figure 3.18 Relative portions of the different groups of flocs and the values of the coefficient of aggregation at given sampling sites throughout a DWTP treatment chain (sites 1–10 were located within a flocculation canal; site 11 was at the entrance to the filters; sites F1 and F32 were the selected filters). Flocculation tests were conducted in the spring at flow rates of (a) 1.5 m³ s⁻¹ and (b) 3.5 m³ s⁻¹.

micro-flocs was relatively higher compared to the winter season, presumably due to the anticipated changes in raw water composition affecting the floc properties. At the higher flow rate (Figure 3.18(b)), the share of micro-flocs was even higher, and they prevailed at the end of the flocculation canal; however, a decrease in the P_{MI} and a relative increase in the P_{PR} again occurred at the entrance to the filters, and primary flocs therefore comprised the majority at the filtration stage. In general, although there were differences in the floc size distributions between the flow rates as well as between the seasons, the efficiency of aggregation represented by $\alpha_{A(Al)}$ was always satisfactory, and flocs suitable for direct sand filtration were formed.

doi: 10.2166/9781789062694_0037

Conclusion

Coagulation–flocculation is a very important step in drinking water treatment owing to its capability of removing many undesirable impurities from water. However, the best efficiency of this process is conditioned by its convenient operation, and the optimal coagulation–flocculation conditions are influenced by many factors, including the nature of the target impurities and the overall composition of raw water. Conducting jar tests is an essential tool for the investigation and optimisation of coagulation–flocculation. Jar testing enables an investigator to find the optimal chemical parameters, such as the coagulant type and dose and the coagulation pH value. In this regard, it is necessary to emphasise that the addition of most coagulants significantly affects the pH of the water, and joint optimisation of the coagulant dose and the coagulation pH value is therefore appropriate. Additionally, jar tests may also serve as a basis for the optimisation of mixing conditions, such as the velocity gradient and mixing time.

The first part of this handbook provides a brief theoretical background explaining the basic principles of coagulation–flocculation and mainly the importance of specific parameters on the coagulation–flocculation efficiency and on the floc properties. The second part then described the working procedure of conducting jar tests in detail, in which some critical aspects were highlighted. Maintaining the suggested scheme of coagulation–flocculation optimisation procedure is not only highly beneficial for achieving the best possible results but also contributes to the comparability and reproducibility of the test results. Additionally, practical information such as the required equipment for performing the jar tests or the description of a flocculation test applicable to simply assess the character of formed flocs were included, as well as clues for evaluating the obtained results.

This handbook is intended for anyone who wants to perform jar tests accurately, and we believe that it will contribute to improve the understanding and application of the important coagulation–flocculation process.

doi: 10.2166/9781789062694_0039

References

Bache D. H. and Gregory R. (2007). Flocs in Water Treatment. IWA Publishing, London.

Bache D. H. and Gregory R. (2010). Flocs and separation processes in drinking water treatment: a review. *Journal of Water Supply: Research and Technology – AQUA*, **59**(1), 16–30.

Bache D. H. and Rasool E. R. (2001). Characteristics of alumino-humic flocs in relation to DAF performance. *Water Science and Technology*, **43**(8), 203–208.

Bache D. H., Rasool E. R., Moffat D. and McGilligan F. J. (1999). On the strength and character of alumino-humic flocs. *Water Science and Technology*, **40**(9), 81–88.

Bagheri G. H., Bonadonna C., Manzella I. and Vonlanthen P. (2015). On the characterization of size and shape of irregular particles. *Powder Technology*, **270**, 141–153.

Barbot E., Dussouillez P., Bottero J. Y. and Moulin P. (2010). Coagulation of bentonite suspension by polyelectrolytes or ferric chloride: floc breakage and reformation. *Chemical Engineering Journal*, **156**(1), 83–91.

Baresova M., Pivokonsky M., Novotna K., Naceradska J. and Branyik T. (2017). An application of cellular organic matter to coagulation of cyanobacterial cells (*Merismopedia tenuissima*). *Water Research*, **122**, 70–77.

Baresova M., Naceradska J., Novotna K., Cermakova L. and Pivokonsky M. (2020). The impact of preozonation on the coagulation of cellular organic matter produced by *Microcystis aeruginosa* and its toxin degradation. *Journal of Environmental Sciences*, **98**, 124–133.

Bernhardt H., Hoyer O., Schell H. and Lusse B. (1985). Reaction mechanisms involved in the influence of algogenic matter on flocculation. *Zeitschrift für Wasser- und Abwasser Forschung*, **18**(1), 18–30.

Birdi K. S. (2003). Handbook of Surface and Colloid Chemistry, 2nd edn. CRC Press, Boca Raton, London, New York, Washington, D.C.

Bouyer D., Liné A. and Do-Quang Z. (2004). Experimental analysis of floc size distribution under different hydrodynamics in a mixing tank. *AIChE Journal*, **50**(9), 2064–2081.

Bouyer D., Coufort C., Liné A. and Do-Quang Z. (2005). Experimental analysis of floc size distributions in a 1-L jar under different hydrodynamics and physicochemical conditions. *Journal of Colloid and Interface Science*, **292**(2), 413–428.

Bratby J. (2006). Coagulation and Flocculation in Water and Wastewater Treatment, 2nd edn. IWA Publishing, London, Seattle.

Bubakova P. and Pivokonsky M. (2012). The influence of velocity gradient on properties and filterability of suspension formed during water treatment. *Separation and Purification Technology*, **92**, 161–167.

Bubakova P., Pivokonsky M. and Pivokonska L. (2011). A method for evaluation of suspension quality easy applicable to practice: the effect of mixing on floc properties. *Journal of Hydrology and Hydromechanics*, **59**(3), 184–195.

Bubakova P., Pivokonsky M. and Filip P. (2013). Effect of shear rate on aggregate size and structure in the process of aggregation and at steady state. *Powder Technology*, **235**, 540–549.

Camp T. R. and Stein P. C. (1943). Velocity gradient and internal work in fluid motion. *Journal of Boston Society of Civil Engineering*, **30**(4), 219–237.

Cao B., Gao B., Liu X., Wang M., Yang Z. and Yue Q. (2011). The impact of pH on floc structure characteristic of polyferric chloride in a low DOC and high alkalinity surface water treatment. *Water Research*, **45**(18), 6181–6188.

Chekli L., Eripret C., Park S. H., Tabatabai S. A. A., Vronska O., Tamburic B., Kim J. H. and Shon H. K. (2017a). Coagulation performance and floc characteristics of polytitanium tetrachloride (PTC) compared with titanium tetrachloride (TiCl$_4$) and ferric chloride (FeCl$_3$) in algal turbid water. *Separation and Purification Technology*, **175**, 99–106.

Chekli L., Corjon E., Tabatabai S. A. A., Naidu G., Tamburic B., Park S. H. and Shon H. K. (2017b). Performance of titanium salts compared to conventional FeCl$_3$ for the removal of algal organic matter (AOM) in synthetic seawater: coagulation performance, organic fraction removal and floc characteristics. *Journal of Environmental Management*, **201**, 28–36.

Cheng W. P. (2002). Comparison of hydrolysis/coagulation behaviour of polymeric and monomeric iron coagulants in humic acid solution. *Chemosphere*, **47**(9), 963–969.

Ching H. W., Tanaka T. S. and Elimelech M. (1994). Dynamics of coagulation of kaolin particles with ferric chloride. *Water Research*, **28**(3), 559–569.

Coufort C., Bouyer D. and Liné A. (2005). Flocculation related to local hydrodynamics in Taylor–Couette reactor and in a jar. *Chemical Engineering Science*, **60**(8–9), 2179–2192.

Coufort C., Bouyer D., Liné A. and Haut B. (2007). Modelling of flocculation using a population balance equation. *Chemical Engineering and Processing*, **46**(12), 1264–1273.

Council Directive 98/83/EC of 3 November 1998 on the quality of water intended for human consumption, *Official Journal of the European Communities* L 330, 05/12/1998 P. 0032-0054.

Creighton T. E. (1993). Proteins: Structures and Molecular Properties, 2nd edn. W.H. Freeman and Company, New York, USA.

De Boer G. B. J., Hoedemakers G. F. M. and Thoenes D. (1989). Coagulation in turbulent flow. Part II. *Chemical Engineering Research and Design*, **67**(3), 308–315.

DiPrima R. C., Eagles P. M. and Ng B. S. (1984). The effect of radius ratio on the stability of Couette flow and Taylor vortex flow. *Physics of Fluids*, **27**(10), 2403–2411.

Duan J. and Gregory J. (2003). Coagulation by hydrolysing metal salts. *Advances in Colloid and Interface Science*, **100–102**, 475–502.

Duan J., Graham N. J. D. and Wilson F. (2003). Coagulation of humic acid by ferric chloride in saline (marine) water conditions. *Water Science and Technology*, **47**(1), 41–48.

Edzwald J. K. (1995). Principles and applications of dissolved air flotation. *Water Science and Technology*, **31**(3–4), 1–23.

Edzwald J. K. (2010). Dissolved air flotation and me. *Water Research*, **44**, 2077–2106.

Ehrl L., Soos M., Morbidelli M. and Bäbler M. U. (2009). Dependence of initial cluster aggregation kinetics on shear rate for particles of different sizes under turbulence. *AIChE Journal*, **55**(12), 3076–3087.

Faust S. D. and Aly O. M. (1998). Chemistry of Water Treatment, 2nd edn. CRC Press, Boca Raton, FL, USA.

Filipenska M., Vasatova P., Pivokonska L., Cermakova L., Gonzales-Torres A., Henderson R. K., Naceradska J. and Pivokonsky M. (2019). Influence of COM-peptides/proteins on the properties of flocs formed at different shear rates. *Journal of Environmental Sciences*, **80**, 116–127.

Fitzpatrick C. S. B., Fradin E. and Gregory J. (2004). Temperature effects on flocculation, using different coagulants. *Journal of Water Supply: Research and Technology – Aqua*, **50**(12), 171–175.

Francois R. J. (1987). Strength of aluminium hydroxide flocs. *Water Research*, **21**(9), 1023–1030.

Gao B. Y., Abbt-Braun G. and Frimmel F. H. (2006). Preparation and evaluation of polyaluminum chloride sulfate (PACS) as a coagulant to remove natural organic matter from water. *Acta Hydrochimica et Hydrobiologica*, **34**, 491–497.

Gonzalez-Torres A., Putnam J., Jefferson B., Stuetz R. M. and Henderson R. K. (2014). Examination of the physical properties of *Microcystis aeruginosa* flocs produced on coagulation with metal salts. *Water Research*, **60**, 197–209.

Gonzalez-Torres A., Rich A. M., Marjo C. E. and Henderson R. K. (2017). Evaluation of biochemical algal floc properties using reflectance Fourier-transform infrared imaging. *Algal Research*, **27**, 345–355.

Gorczyca B. and Ganczarczyk J. (1999). Structure and porosity of alum coagulation flocs. *Water Quality Research Journal of Canada*, **34**(4), 653–666.

Gregory J. (2006). Particles in Water: Properties and Processes. CRC Press, Boca Raton, London, New York.

Gregory J. and Dupont V. (2001). Properties of flocs produced by water treatment coagulants. *Water Science and Technology*, **44**(10), 231–236.

Gregory J. and O'Melia C. R. (1989). Fundamentals of flocculation. *Critical Reviews in Environmental Control*, **19**(3), 185–230.

Gregory J. and Yukselen M. A. (2001). Break-up and re-formation of flocs formed by hydrolyzing coagulants and polymeric flocculants. In: Chemical Water and Wastewater Treatment, H. Hahn, E. Hoffmann and H. Odegaard (eds), IWA Publishing, London, pp. 29–38.

Han M. and Lawler D. F. (1991). Interaction of two settling rates and collision efficiencies. *Journal of the Hydraulics Division – ASCE*, **17**(10), 1296–1289.

He W., Xue L., Gorczyca B., Nan J. and Shi Z. (2018a). Experimental and CFD studies of floc growth dependence on baffle width in square stirred-tank reactors for flocculation. *Separation and Purification Technology*, **190**, 228–242.

He W., Zhao Z., Nan J., Xie Z. and Lu W. (2018b). The role of mixing hydrodynamics on floc growth in unbaffled square stirred-tank reactors for flocculation. *Journal of Environmental Chemical Engineering*, **6**(2), 3041–3053.

He W., Lu W., Xu S., Huang M. and Li H. (2019). Comparative analysis on floc morphological evolution in cylindrical and square stirred-tank flocculating reactors with or without baffles: flocculation-test and CFD-aided investigations. *Chemical Engineering Research and Design*, **147**, 278–291.

Heath A. R. and Koh P. T. L. (2003). Combined population balance and CFD modelling of particle aggregation by polymeric flocculant. Proc. Third International Conference on CFD in the Minerals and Process Industries, CSIRO, Melbourne, Australia, pp. 339–344.

Henderson R., Sharp E., Jarvis P., Parsons S. and Jefferson B. (2006). Identifying the linkage between particle characteristics and understanding coagulation performance. *Water Science and Technology: Water Supply*, **6**(1), 31–38.

Henderson R., Parsons S. A. and Jefferson B. (2008a) The impact of algal properties and preoxidation on solid-liquid separation of algae. *Water Research*, **42**(8–9), 1827–1845.

Henderson R. K., Baker A., Parsons S. A. and Jefferson B. (2008b). Characterisation of algogenic organic matter extracted from cyanobacteria, green algae and diatoms. *Water Research*, **42**(13), 3435–3445.

Henderson R. K., Parsons S. A. and Jefferson B. (2010). The impact of differing cell and algogenic organic matter (AOM) characteristics on the coagulation and flotation of algae. *Water Research*, **44**(12), 3617–3624.

Hereit F., Mutl S. and Vagner V. (1980). Direct measurement of floc breakage in flowing suspension. *Journal of Water Supply: Research and Technology – AQUA*, **29**, 95–99.

Hopkins D. C. and Ducoste J. J. (2003). Characterizing flocculation under heterogeneous turbulence. *Journal of Colloid and Interface Science*, **264**(1), 184–194.

Hunter R. J. (2001). Foundations of Colloid Science. Oxford University Press, Oxford.

Hussain S., van Leeuwen J., Chow C. W. K., Aryal R., Beecham S., Duan J. and Drikas M. (2014). Comparison of the coagulation performance of tetravalent titanium and zirconium salts with alum. *Chemical Engineering Journal*, **254**, 635–646.

Ives K. J. (1978). The Scientific Basis of Flocculation. Sijthoff and Noordhoff, Netherlands.

Jarvis P., Jefferson B. and Parsons S. A. (2005a). Breakage, regrowth, and fractal nature of natural organic matter flocs. *Environmental Science and Technology*, **39**(7), 2307–2314.

Jarvis P., Jefferson B., Gregory J. and Parsons S. A. (2005b). A review of floc strength and breakage. *Water Research*, **39**(14), 3121–3137.

Jarvis P., Sharp E., Pidou M., Molinder R., Parsons S. A. and Jefferson B. (2012). Comparison of coagulation performance and floc properties using a novel zirconium coagulant against traditional ferric and alum coagulants. *Water Research*, **46**(13), 4179–4187.

Jiang J.-Q. and Graham N. J. D. (1998). Preparation and characterisation of an optimal polyferric sulphate (PFS) as a coagulant for water treatment. *Journal of Chemical Technology and Biotechnology*, **73**, 351–358.

Jiao R., Fabris R., Chow C. W. K., Drikas M., van Leeuwen J. and Wang D. (2016). Roles of coagulant species and mechanisms on floc characteristics and filterability. *Chemosphere*, **150**, 211–218.

Kataoka K. (1986). Taylor vortices and instabilities in circular Couette flows. In: Encyclopaedia of Fluid Mechanics, N. P. Cheremisinoff (ed.), Gulf Publishing, Houston, Vol. 1, pp. 237–273.

Khiadani M., Kolivand R., Ahooghalandari M. and Mohaje M. (2013). Removal of turbidity from water by dissolved air flotation and conventional sedimentation systems using poly aluminium chloride as coagulant. *Desalination and Water Treatment*, **52**(4–6), 985–989.

Kim J. S. and Kang L. S. (1998). Investigation of coagulation mechanisms with Fe(III) salt using jar tests and flocculation dynamics. *Environmental Engineering Research*, **3**(1), 11–19.

Kong Y., Ma Y., Ding L., Ma J., Zhang H., Chen Z. and Shen J. (2021). Coagulation behaviors of aluminum salts towards humic acid: detailed analysis of aluminum speciation and transformation. *Separation and Purification Technology*, **259**, 1181737.

Kusters K. A., Wijers J. G. and Thoenes D. (1997). Aggregation kinetics of small particles in agitated vessels. *Chemical Engineering Science*, **52**(1), 107–121.

Lei G., Ma J., Guan X., Song A. and Cui Y. (2009). Effect of basicity on coagulation performance of polyferric chloride applied in eutrophicated raw water. *Desalination*, **247**, 518–529.

Li R., Gao B., Huang X., Dong H., Li X., Yue Q., Wang Y. and Li Q. (2014). Compound bioflocculant and polyaluminum chloride in kaolin-humic acid coagulation: factors influencing coagulation performance and floc characteristics. *Bioresource Technology*, **172**, 8–15.

Li T., Zhu Z., Wang D., Yao C. and Tang H. (2006). Characterization of floc size, strength and structure under various coagulation mechanisms. *Powder Technology*, **168**(2), 104–110.

Liang L., Peng Y., Tan J. and Xie G. (2015). A review of the modern characterization techniques for flocs in mineral processing. *Minerals Engineering* **84**, 130–144.

Lin J.-L., Huang C., Dempsey B. and Hu J.-Y. (2014). Fate of hydrolyzed Al species in humic acid coagulation. *Water Research*, **56**, 314–324.

Liu H., Hu C., Zhao H. and Qu J. (2009). Coagulation of humic acid by PACl with high content of Al_{13}: the role of aluminum speciation. *Separation and Purification Technology*, **70**(2), 225–230.

Lu X. Q., Chen Z. and Yang X. (1999). Spectroscopic study of aluminium speciation in removing humic substances by Al coagulation. *Water Research*, **33**(15), 3271–3280.

Merkus H. G. (2009). Particle Size Measurements: Fundamentals, Practice, Quality. Springer, London.

Morris J. K. and Knocke W. R. (1984). Temperature effects on the use of metal-ion coagulants for water treatment. *Journal of American Water Works Association*, **76**(3), 74–79.

Morrow J. J. and Rausch E. G. (1974). Colloid destabilization with cationic polyelectrolytes as affected by velocity gradients. *Journal of American Water Works Association*, **66**(11), 646–653.

Moruzzi R. B., de Oliveira A. L., da Conceição F. T., Gregory J. and Campos L. C. (2017). Fractal dimension of large aggregates under different flocculation conditions. *Science of the Total Environment*, **609**, 807–814.

Mutl S., Knesl B. and Polasek P. (1999). Application of a fluidized layer of granular material in the treatment of surface water. Part 1: aggregation efficiency of the layer. *Journal of Water Supply: Research and Technology – AQUA*, **48**(1), 24–30.

Mutl S., Polasek P., Pivokonsky M. and Kloucek O. (2006). The influence of G and T on the course of aggregation at the treatment of medium polluted surface water. *Water Science and Technology: Water Supply*, **6**, 39–48.

Naceradska J., Novotna K., Cermakova L., Cajthaml T. and Pivokonsky M. (2019a). Investigating the coagulation of nonproteinaceous algal organic matter: optimization of coagulation performance and identification of removal mechanisms. *Journal of Environmental Sciences*, **79**, 25–34.

Naceradska J., Pivokonska L. and Pivokonsky M. (2019b). On the importance of pH value in coagulation. *Journal of Water Supply: Research and Technology – AQUA*, **68**(3), 222–230.

Nan J., Wang Z., Yao M., Yang Y. and Zhang X. (2016). Characterization of re-grown floc size and structure: effect of mixing conditions during floc growth, breakage and re-growth process. *Environmental Science and Pollution Research*, **23**(23), 23750–23757.

Newcombe G., Drikas M., Assemi S. and Beckett R. (1997). Influence of characterised natural organic material on activated carbon adsorption: I. Characterisation of concentrated reservoir water. *Water Research*, **31**(5), 965–972.

Ngo H. H., Vigneswaran P. and Dhamappa H. B. (1995). Optimization of direct filtration: experiments and mathematical models. *Environmental Technology*, **16**(1), 55–63.

Oles V. (1992). Shear-induced aggregation and break-up of polystyrene latex particles. *Journal of Colloid and Interface Science*, **154**(2), 351–358.

Paralkar A. and Edzwald J. K. (1996). Effect of ozone on EOM and coagulation. *Journal of the American Water Works Association*, **88**(4), 143–154.

Parker D. S., Kaufman W. J. and Jenkins D. (1972). Floc breakup in turbulent flocculation processes. *Journal of the Sanitary Engineering Division*, **98**(1), 79–99.

Pernitsky D. J. and Edzwald J. K. (2003). Solubility of polyaluminium coagulants. *Journal of Water Supply: Research and Technology – Aqua*, **52**(6), 395–406.

Persson I. (2010). Hydrated metal ions in aqueous solution: how regular are their structures? *Pure and Applied Chemistry*, **82**, 1901–1917.

Pivokonsky M., Pivokonska L. and Tomaskova H. (2008). Aggregation capability of a fluidised layer of granular material during treatment of water with high DOC and low alkalinity. *Water Science and Technology: Water Supply*, **8**(1), 9–17.

Pivokonsky M., Pivokonska L., Baumeltova J. and Bubakova P. (2009a). The effect of cellular organic matter produced by cyanobacteria *Microcystis aeruginosa* on water purification. *Journal of Hydrology and Hydromechanics*, **57**(2), 121–129.

Pivokonsky M., Polasek P., Pivokonska L. and Tomaskova H. (2009b). Optimized reaction conditions for removal of cellular organic matter of *Microcystis aeruginosa* during the destabilization and aggregation process using ferric sulfate in water purification. *Water Environment Research*, **81**(5), 514–522.

Pivokonsky M., Bubakova P., Pivokonska L. and Hnatukova P. (2011). The effect of global velocity gradient on the character and filterability of aggregates formed during the coagulation/flocculation process. *Environmental Technology*, **32**(12), 1355–1366.

Pivokonsky M., Safarikova J., Bubakova P. and Pivokonska L. (2012). Coagulation of peptides and proteins produced by *Microcystis aeruginosa*: interaction mechanisms and the effect of Fe-peptide/protein complexes formation. *Water Research*, **46**, 5583–5590.

Pivokonsky M., Naceradska J., Brabenec T., Novotna K., Baresova M. and Janda V. (2015) The impact of interactions between algal organic matter and humic substances on coagulation. *Water Research*, **84**, 278–285.

Polasek P. (2011). Influence of velocity gradient on optimisation of the aggregation process and physical properties of formed aggregates: part 1. Inline high-density suspension (IHDS) aggregation process. *Journal of Hydrology and Hydromechanics*, **59**(2), 107–117.

Polasek P. and Mutl S. (1995). Guidelines to Coagulation and Flocculation for Surface Waters: Volume 1 – Design Principles for Coagulation and Flocculation Systems. Polasek and Associates, Marshalltown, South Africa.

Polasek P. and Mutl S. (2005). Optimisation of reaction conditions of particle aggregation in water purification – back to basics. *Water SA*, **31**(1), 61–72.

Prat O. P. and Ducoste J. J. (2006). Modeling spatial distribution of floc size in turbulent processes using the quadrature method of moment and computational fluid dynamics. *Chemical Engineering Science*, **61**(1), 75–86.

Rossini M., Garrido J. G. and Galluzzo M. (1999). Optimatization of the coagulation/flocculation treatment: influence of rapid mix parameters. *Water Research*, **33**(8), 1817–1826.

Safarikova J., Baresova M., Pivokonsky M. and Kopecka I. (2013). Influence of peptides and proteins produced by cyanobacterium *Microcystis aeruginosa* on the coagulation of turbid waters. *Separation and Purification Technology*, **118**, 49–57.

Selomulya C., Amal R., Bushell G. and Waite T. D. (2001). Evidence of shear rate dependence on restructuring and breakup of latex aggregates. *Journal of Colloid and Interface Science*, **236**(1), 67–77.

Selomulya C., Bushell G., Amal R. and Waite T. D. (2003). Understanding the role of restructuring in flocculation: the application of a population balance model. *Chemical Engineering Science*, **58**(2), 327–338.

Serra T., Colomer J. and Logan B. E. (2008). Efficiency of different shear devices on flocculation. *Water Research*, **42**(4–5), 1113–1121.

Shi B., Wei Q., Wang D., Zhu Z. and Tang H. (2007). Coagulation of humic acid: the performance of preformed and non-preformed Al species. *Colloids and Surfaces A: Physicochemical and Engineering Aspects*, **296**(1–3), 141–148.

Smoluchowski M. (1916). Drei Vorträge über Diffusion, Brownsche Bewegung und Koagulation von Kolloidteilchen (Three lectures on diffusion, Brownian motion and coagulation of colloidal particles). *Zeitschrift für Physikalische Chemie*, **17**, 557–585.

Smoluchowski M. (1917). Versuch einer mathematischen Theorie der Koagulationskinetik kolloider Lösungen (Testing a mathematical theory of the coagulation kinetics of colloidal solutions). *Zeitschrift für Physikalische Chemie*, **92**, 124–168.

Spicer P. T. and Pratsinis S. E. (1996a). Shear-induced flocculation: the evolution of floc structure and the shape of the size distribution at steady state. *Water Research*, **30**(5), 1049–1056.

Spicer P. T. and Pratsinis S. E. (1996b). Coagulation and fragmentation: universal steady-state particle-size distribution. *AIChE Journal*, **42**(6), 1612–1620.

Spicer P. T., Keller W. and Pratsinis S. E. (1996). The effect of impeller type on floc size and structure during shear-induced flocculation. *Journal of Colloid and Interface Science*, **184**(1), 112–122.

Spicer P. T., Pratsinis S. E., Raper J., Amal R., Bushell G. and Meesters G. (1998). Effect of shear schedule on particle size, density and structure during flocculation in stirred tanks. *Powder Technology*, **97**(1), 26–34.

Stumm W. and Morgan J. J. (1996). Aquatic Chemistry, 3rd edn. John Wiley & Sons, New York.

Tambo N. and Watanabe Y. (1979). Physical characterstics of flocs. I. The floc density function and aluminium floc. *Water Research*, **13**, 409–419.

Valade M. T., Edzwald J. K., Tobiason J. E., Dahlquist J., Hedberg T. and Amato T. (1996). Particle removal by flotation and filtration: pretreatment effects. *Journal of American Water Works Association*, **88**(12), 35–46.

van Benschoten J. E. and Edzwald J. K. (1990). Chemical aspects of coagulation using aluminium salts – I. Hydrolytic reactions of alum and polyaluminium chloride. *Water Research*, **24**(12), 1519–1526.

Vandamme D., Muylaert K., Fraeye I. and Foubert I. (2014). Floc characteristics of *Chlorella vulgaris*: influence of flocculation mode and presence of organic matter. *Bioresource Technology*, **151**, 383–387.

Vasatova P., Filipenska M., Petricek R. and Pivokonsky M. (2020). On the importance of mixing characterization and application in the water treatment process. *Journal of Water Supply: Research and Technology-Aqua*, **69**(6), 639–646.

Wang D., Sun W., Xu Y., Tang H. and Gregory J. (2004). Speciation stability of inorganic polymer flocculant PACl. *Colloids and Surfaces A: Physicochemical and Engineering Aspects*, **243**, 1–10.

Wang L. K., Hung Y.-T. and Shammas N. K. (2005). Physicochemical Treatment Processes. Humana Press, Totowa, New Jersey.

Wang Y., Gao B. Y., Xu X. M., Xu W. Y. and Xu G. Y. (2009). Characterization of floc size, strength and structure in various aluminum coagulants treatment. *Journal of Colloid and Interface Science*, **332**, 354–359.

Wang J., Xu W., Xu J., Wei D., Feng H. and Xu Z. (2016). Effect of aluminum speciation and pH on in-line coagulation/diatomite microfiltration process: correlations between aggregate characteristics and membrane fouling. *Journal of Molecular Liquids*, **224**, 492–501.

Wang B., Shui Y., He M. and Liu P. (2017). Comparison of flocs characteristics using before and after composite coagulants under different coagulation mechanisms. *Biochemical Engineering Journal*, **121**, 107–117.

Widrig D. L., Gray K. A. and McAuliffe K. S. (1996). Removal of algal-derived organic material by preozonation and coagulation: monitoring changes in organic quality by pyrolysis-GC-MS. *Water Research*, **30**(11), 2621–2632

Williams R. A., Peng S. J. and Naylor A. (1992). In situ measurement of particle aggregation and breakage kinetics in a concentrated suspension. *Powder Technology*, **73**(1), 75–83.

Xiao F., Zhang B. and Lee C. (2008). Effects of low temperature on aluminum(III) hydrolysis: theoretical and experimental studies. *Journal of Environmental Sciences*, **20**, 907–914.

Xu W., Gao B., Yue Q. and Wang Y. (2010). Effect of shear force and solution pH on flocs breakage and re-growth formed by nano-Al$_{13}$ polymer. *Water Research*, **44**(6), 1893–1899.

Zand A. D. and Hoveidi H. (2015). Comparing aluminium sulfate and poly-aluminium chloride (PAC) performance in turbidity removal from synthetic water. *Journal of Applied Biotechnology Reports*, **2**(3), 287–292.

Zhang Z., Jing R., He S., Qian J., Zhang K., Ma G., Chang X., Zhang M. and Li Y. (2018). Coagulation of low temperature and low turbidity water: adjusting basicity of polyaluminum chloride (PAC) and using chitosan as coagulant aid. *Separation and Purification Technology*, **206**, 131–139.

Zhao Y., Gao B., Cao B., Yang Z., Yue Q., Shon H. and Kim J.-H. (2011a). Comparison of coagulation behavior and floc characteristics of titanium tetrachloride (TiCl$_4$) and polyaluminum chloride (PACl) with surface water treatment. *Chemical Engineering Journal*, **166**, 544–550.

Zhao Y. X., Gao B. Y., Shon H. K., Cao B. C. and Kim J.-H. (2011b). Coagulation characteristics of titanium (Ti) salt coagulant compared with aluminum (Al) and iron (Fe) salts. *Journal of Hazardous Materials*, **185**(2–3), 1536–1542.

Zhao Y. X., Gao B. Y., Zhang G. Z., Phuntsho S., Wang Y., Yue Q. Y., Li Q. and Shon H. K. (2013a). Comparative study of floc characteristics with titanium tetrachloride against conventional coagulants: effect of coagulant dose, solution pH, shear force and break-up period. *Chemical Engineering Journal*, **233**, 70–79.

Zhao Y. X., Phuntsho S., Gao B. Y., Huang X., Qi Q. B., Yue Q. Y., Wang Y., Kim J.-H. and Shon H. K. (2013b). Preparation and characterization of novel polytitanium tetrachloride coagulant for water purification. *Environmental Science and Technology*, **47**(22), 12966–12975.

Zhao Y. X., Phuntsho S., Gao B. Y., Yang Y. Z., Kim J.-H. and Shon H. K. (2015). Comparison of a novel polytitanium chloride coagulant with polyaluminium chloride: coagulation performance and floc characteristics. *Journal of Environmental Management*, **147**, 194–202.

Zhao Z., Sun W., Ray A. K., Mao T. and Ray M. B. (2020). Coagulation and disinfection by-products formation potential of extracellular and intracellular matter of algae and cyanobacteria. *Chemosphere*, **245**, 125669.

doi: 10.2166/9781789062694_0045

Symbols and Abbreviations

α_A	dimensionless	degree of aggregation
α_D	dimensionless	degree of destabilisation
γ	dimensionless	constant in equation (2.1)
η	Pa s	dynamic viscosity
ν	$m^2\,s^{-1}$	kinematic viscosity
ρ	$mg\,L^{-1}$	mass concentration
ρ_f	$kg\,m^{-3}$	fluid density
ρ_v	$kg\,m^{-3}$	water density
ω	$rad\,s^{-1}$	angular velocity
A_1–A_6	dimensionless	empirical constants referring to the specific mixer dimensions and geometry
A_r	dimensionless	relative atomic mass
c	$mmol\,L^{-1}$	molar concentration
C	dimensionless	constant in equation (2.1)
C_0	$mg\,L^{-1}$	initial concentration of a given parameter (at the beginning of settling)
C_5	$mg\,L^{-1}$	concentration after 5 min of settling
C_{60}	$mg\,L^{-1}$	concentration after 60 min of settling
$C_{F(60)}$	$mg\,L^{-1}$	concentration after 60 min of settling and subsequent centrifugation
$C_{F(A)}$	$mg\,L^{-1}$	concentration in a centrifuged sample collected after flocculation mixing
$C_{F(HM)}$	$mg\,L^{-1}$	concentration in a centrifuged sample collected after homogenisation mixing
$d/d_{av}/d_{max}$	μm	floc diameter/average floc diameter/maximum floc diameter
D	$mg\,L^{-1}$	coagulant dose
d_m	m	diameter of a stirrer (mixer)
d_N	m	diameter of a mixing vessel
f	min^{-1}	rotation frequency
g	$m\,s^{-2}$	gravitational acceleration
$\bar{G}$	s^{-1}	mean velocity gradient (global shear rate)
$\bar{G}_{RM}$	s^{-1}	mean velocity gradient of rapid flocculation mixing
$\bar{G}_{SM}$	s^{-1}	mean velocity gradient of slow flocculation mixing
h_d	m	distance of a mixer (stirrer) from the bottom of a mixing vessel
h_m	m	height of a stirrer (mixer)
h_N	m	height of a mixing vessel filled with water (from the bottom to the water surface)
M (Section 3.3)	N m	torque
M (Appendix)	$g\,mol^{-1}$	molar mass
M_r	dimensionless	relative molecular mass
P	W	power input
Po	dimensionless	power number

Re	dimensionless	Reynolds number
R_1	m	diameter of an inner cylinder of a Taylor–Couette reactor
R_2	m	diameter of an outer cylinder of a Taylor–Couette reactor
s_m	m	width of a stirrer (mixer)
t	s	mixing time
T	°C	temperature
t_{HM}	min	time of homogenisation mixing
t_{RM}	min	time of rapid flocculation mixing
t_{SM}	min	time of slow flocculation mixing
U	m s^{-1}	mean flow velocity
v	m s^{-1}	flow velocity of water
V	dm^3	volume
w	dimensionless	mass fraction
$ANC_{4.5}$	acid neutralising capacity to pH $= 4.5$	
AOM	algal organic matter	
COD_{Mn}	chemical oxygen demand by potassium permanganate	
COM	cellular organic matter	
DOC	dissolved organic carbon	
DWTP	drinking water treatment plant	
HS	humic substances	
HM	homogenisation mixing	
MA	macro-flocs	
MI	micro-flocs	
NF	non-flocculated portion	
NOM	natural organic matter	
P_{MI}	portion of micro-flocs	
P_{MA}	portion of macro-flocs	
P_{NF}	non-flocculated portion	
P_{PR}	portion of primary flocs	
PACl	polyaluminium chloride	
PFS	polyferric sulphate	
PR	primary flocs	
RM	rapid flocculation mixing	
SM	slow flocculation mixing	
TOC	total organic carbon	
UV_{254}	ultra-violet absorbance at the wavelength of 254 nm	

doi: 10.2166/9781789062694_0047

Appendices

APPENDIX 1: PHYSICAL PROPERTIES OF WATER

Table 1A Dynamic viscosity η, density ρ_v and kinematic viscosity ν of water depending on temperature T at an atmospheric pressure of 101.325 kPa ($\eta = \rho_v \cdot \nu$).

T (°C)	η (10^{-3} Pa s^{-1})	ρ_v (kg m^{-3})	ν (m^2 s^{-1})
0	1.792	999.82	1.7923
1	1.731	999.89	1.7312
2	1.674	999.94	1.6741
3	1.620	999.98	1.6200
4	1.569	1000.00	1.5690
5	1.520	1000.00	1.5200
6	1.473	999.99	1.4730
7	1.429	999.96	1.4291
8	1.386	999.91	1.3861
9	1.346	999.85	1.3462
10	1.308	999.77	1.3083
11	1.271	999.68	1.2714
12	1.236	999.58	1.2365
13	1.202	999.46	1.2026
14	1.170	999.33	1.1708
15	1.139	999.19	1.1399
16	1.109	999.03	1.1101
17	1.081	998.86	1.0822
18	1.054	998.68	1.0554
19	1.028	998.49	1.0296
20	1.003	998.29	1.0047
21	0.979	998.08	0.9809
22	0.955	997.86	0.9570
23	0.933	997.62	0.9352
24	0.911	997.38	0.9134
25	0.891	997.13	0.8936

APPENDIX 2: MOLECULAR (ATOMIC) MASSES

Table 2A Relative molecular (M_r) or atomic (A_r) masses of the most commonly used compounds and elements in water technology.

Compound/Element	Relative Molecular/Atomic Mass
Al	26.9815
$AlCl_3$	133.3405
$AlCl_3 \cdot 6H_2O$	241.4317
$Al_2(SO_4)_3$	342.1538
$Al_2(SO_4)_3 \cdot 18H_2O$	666.4274
$Al(OH)_3$	78.0034
Al_2O_3	101.9612
Fe	55.8470
$FeCl_3$	162.2060
$FeCl_3 \cdot 6H_2O$	270.2972
$FeSO_4$	151.9106
$Fe_2(SO_4)_3$	399.8848
$Fe_2(SO_4)_3 \cdot 9H_2O$	562.0216
$FeClSO_4$	187.3636
$FeClSO_4 \cdot 6H_2O$	295.4548
$Fe(OH)_3$	106.8689
Fe_2O_3	159.6922
Ca	40.0780
CaO	56.0774
$Ca(OH)_2$	74.0926
H_2SO_4	98.0794

APPENDIX 3: CONVERSION OF MASS CONCENTRATIONS TO MOLAR CONCENTRATIONS (AND VICE VERSA) OF SEVERAL COAGULANTS

Table 3A Conversion of the mass concentration ρ (Al$_2$(SO$_4$)$_3$) to the Al content and the molar concentration c.

ρ (Al$_2$(SO$_4$)$_3$) (mg L^{-1})	ρ (Al) (mg L^{-1})	c (Al$_2$(SO$_4$)$_3$) (mmol L^{-1})
1	0.1577	0.0029
5	0.7886	0.0146
10	1.5772	0.0292
15	2.3658	0.0438
20	3.1543	0.0585
25	3.9429	0.0731
30	4.7315	0.0877
35	5.5201	0.1023
40	6.3087	0.1169
45	7.0973	0.1315
50	7.8859	0.1461
55	8.6744	0.1607
60	9.4630	0.1754
65	10.2516	0.1900
70	11.0402	0.2046
75	11.8288	0.2192
80	12.6174	0.2338
85	13.4060	0.2484
90	14.1945	0.2630
95	14.9831	0.2777
100	15.7717	0.2923
105	16.5603	0.3069
110	17.3489	0.3215
115	18.1375	0.3361
120	18.9261	0.3507
125	19.7146	0.3653
130	20.5032	0.3799
135	21.2918	0.3946
140	22.0804	0.4092
145	22.8690	0.4238
150	23.6576	0.4384

$$c\left(\text{Al}_2\left(\text{SO}_4\right)_3\right) = \frac{\rho\left(\text{Al}_2\left(\text{SO}_4\right)_3\right)}{M\left(\text{Al}_2\left(\text{SO}_4\right)_3\right)}$$

$$c\left(\text{Al}\right) = 2 \cdot \rho\left(\text{Al}_2(\text{SO}_4)_3\right) \cdot \frac{M\left(\text{Al}\right)}{M\left(\text{Al}_2\left(\text{SO}_4\right)_3\right)}$$

Table 3B Conversion of the mass concentration ρ (Al$_2$(SO$_4$)$_3 \cdot$ 18H$_2$O) to the Al content and the molar concentration c.

ρ (Al$_2$(SO$_4$)$_3 \cdot$ 18 H$_2$O) (mg L^{-1})	ρ (Al) (mg L^{-1})	c (Al$_2$(SO$_4$)$_3 \cdot$ 18 H$_2$O) (mmol L^{-1})
1	0.0810	0.0015
5	0.4049	0.0075
10	0.8097	0.0150
15	1.2146	0.0225
20	1.6195	0.0300
25	2.0243	0.0375
30	2.4292	0.0450
35	2.8341	0.0525
40	3.2390	0.0600
45	3.6438	0.0675
50	4.0487	0.0750
55	4.4536	0.0825
60	4.8584	0.0900
65	5.2633	0.0975
70	5.6682	0.1050
75	6.0730	0.1125
80	6.4779	0.1200
85	6.8828	0.1275
90	7.2876	0.1350
95	7.6925	0.1426
100	8.0974	0.1501
105	8.5023	0.1576
110	8.9071	0.1651
115	9.3120	0.1726
120	9.7169	0.1801
125	10.1217	0.1876
130	10.5266	0.1951
135	10.9315	0.2026
140	11.3363	0.2101
145	11.7412	0.2176
150	12.1461	0.2251

$$c\left(\text{Al}_2\left(\text{SO}_4\right)_3 \cdot 18\text{H}_2\text{O}\right) = \frac{\rho\left(\text{Al}_2\left(\text{SO}_4\right)_3 \cdot 18\text{H}_2\text{O}\right)}{M\left(\text{Al}_2\left(\text{SO}_4\right)_3 \cdot 18\text{H}_2\text{O}\right)}$$

$$c\left(\text{Al}\right) = 2 \cdot \rho\left(\text{Al}_2(\text{SO}_4)_3 \cdot 18\text{H}_2\text{O}\right) \cdot \frac{M\left(\text{Al}\right)}{M\left(\text{Al}_2\left(\text{SO}_4\right)_3 \cdot 18\text{H}_2\text{O}\right)}$$

Table 3C Conversion of the mass concentration ρ (FeCl$_3$) to the Fe content and the molar concentration c.

ρ (FeCl$_3$) (mg L^{-1})	ρ (Fe) (mg L^{-1})	c (FeCl$_3$) (mmol L^{-1})
1	0.344	0.0062
5	1.721	0.0308
10	3.443	0.0617
15	5.164	0.0925
20	6.886	0.1233
25	8.607	0.1541
30	10.329	0.1850
35	12.050	0.2158
40	13.772	0.2466
45	15.493	0.2774
50	17.214	0.3083
55	18.936	0.3391
60	20.657	0.3699
65	22.379	0.4007
70	24.100	0.4316
75	25.822	0.4624
80	27.543	0.4932
85	29.265	0.5240
90	30.986	0.5549
95	32.707	0.5857
100	34.429	0.6165
105	36.150	0.6473
110	37.872	0.6782
115	39.593	0.7090
120	41.315	0.7398
125	43.036	0.7706
130	44.758	0.8015
135	46.479	0.8323
140	48.200	0.8631
145	49.922	0.8939
150	51.643	0.9248

Table 3D Conversion of the mass concentration ρ FeCl$_3 \cdot$6H$_2$O to the Fe content and the molar concentration c.

ρ (FeCl$_3 \cdot$6H$_2$O) (mg L^{-1})	ρ (Fe) (mg L^{-1})	c (FeCl$_3 \cdot$6H$_2$O) (mmol L^{-1})
1	0.207	0.0037
5	1.033	0.0185
10	2.066	0.0370
15	3.099	0.0555
20	4.132	0.0740
25	5.165	0.0925
30	6.198	0.1110
35	7.231	0.1295
40	8.264	0.1480
45	9.297	0.1665
50	10.330	0.1850
55	11.363	0.2035
60	12.396	0.2220
65	13.429	0.2405
70	14.462	0.2590
75	15.496	0.2775
80	16.529	0.2960
85	17.562	0.3145
90	18.595	0.3330
95	19.628	0.3515
100	20.661	0.3700
105	21.694	0.3885
110	22.727	0.4070
115	23.760	0.4255
120	24.793	0.4440
125	25.826	0.4625
130	26.859	0.4810
135	27.892	0.4995
140	28.925	0.5180
145	29.958	0.5364
150	30.991	0.5549

$$c\left(FeCl_3 \cdot 6H_2O\right) = \frac{\rho\left(FeCl_3 \cdot 6H_2O\right)}{M\left(FeCl_3 \cdot 6H_2O\right)}$$

$$c\left(Fe\right) = 2 \cdot \rho\left(FeCl_3 \cdot 6H_2O\right) \cdot \frac{M\left(Fe\right)}{M\left(FeCl_3 \cdot 6H_2O\right)}$$

Table 3E Conversion of the mass concentration ρ Fe$_2$(SO$_4$)$_3$ to the Fe content and the molar concentration c.

ρ (Fe$_2$(SO$_4$)$_3$) (mg L^{-1})	ρ (Fe) (mg L^{-1})	c (Fe$_2$(SO$_4$)$_3$) (mmol L^{-1})
1	0.279	0.0025
5	1.397	0.0125
10	2.793	0.0250
15	4.190	0.0375
20	5.586	0.0500
25	6.983	0.0625
30	8.379	0.0750
35	9.776	0.0875
40	11.172	0.1000
45	12.569	0.1125
50	13.966	0.1250
55	15.362	0.1375
60	16.759	0.1500
65	18.155	0.1625
70	19.552	0.1751
75	20.948	0.1876
80	22.345	0.2001
85	23.741	0.2126
90	25.138	0.2251
95	26.534	0.2376
100	27.931	0.2501
105	29.328	0.2626
110	30.724	0.2751
115	32.121	0.2876
120	33.517	0.3001
125	34.914	0.3126
130	36.310	0.3251
135	37.707	0.3376
140	39.103	0.3501
145	40.500	0.3626
150	41.897	0.3751

$$c\left(\mathrm{Fe}_2\left(\mathrm{SO}_4\right)_3\right) = \frac{\rho\left(\mathrm{Fe}_2\left(\mathrm{SO}_4\right)_3\right)}{M\left(\mathrm{Fe}_2\left(\mathrm{SO}_4\right)_3\right)}$$

$$c\left(\mathrm{Fe}\right) = 2 \cdot \rho\left(\mathrm{Fe}_2(\mathrm{SO}_4)_3\right) \cdot \frac{M\left(\mathrm{Fe}\right)}{M\left(\mathrm{Fe}_2\left(\mathrm{SO}_4\right)_3\right)}$$

Table 3F Conversion of the mass concentration ρ $Fe_2(SO_4)_3 \cdot 9H_2O$ to the Fe content and the molar concentration c.

ρ ($Fe_2(SO_4)_3 \cdot 9H_2O$) (mg L^{-1})	ρ (Fe) (mg L^{-1})	c ($Fe_2(SO_4)_3 \cdot 9H_2O$) (mmol L^{-1})
1	0.199	0.0018
5	0.994	0.0089
10	1.987	0.0178
15	2.981	0.0267
20	3.975	0.0356
25	4.968	0.0445
30	5.962	0.0534
35	6.956	0.0623
40	7.949	0.0712
45	8.943	0.0801
50	9.937	0.0890
55	10.930	0.0979
60	11.924	0.1068
65	12.918	0.1157
70	13.911	0.1246
75	14.905	0.1334
80	15.898	0.1423
85	16.892	0.1512
90	17.886	0.1601
95	18.879	0.1690
100	19.873	0.1779
105	20.867	0.1868
110	21.860	0.1957
115	22.854	0.2046
120	23.848	0.2135
125	24.841	0.2224
130	25.835	0.2313
135	26.829	0.2402
140	27.822	0.2491
145	28.816	0.2580
150	29.810	0.2669

$$c\left(Fe_2\left(SO_4\right)_3 \cdot 9H_2O\right) = \frac{\rho\left(Fe_2\left(SO_4\right)_3 \cdot 9H_2O\right)}{M\left(Fe_2\left(SO_4\right)_3 \cdot 9H_2O\right)}$$

$$c\left(Fe\right) = 2 \cdot \rho\left(Fe_2(SO_4)_3 \cdot 9H_2O\right) \cdot \frac{M\left(Fe\right)}{M\left(Fe_2\left(SO_4\right)_3 \cdot 9H_2O\right)}$$

APPENDIX 4: PREPARATION OF 1% FE$_2$(SO$_4$)$_3$ SOLUTION

Stock solution:

$Fe_2(SO_4)_3 \cdot 9H_2O$
$w = 59.8$ wt%
$\rho = 1.538$ g cm$^{-3} = \rho_1$
Conversion to anhydrous $Fe_2(SO_4)_3$:
$M(Fe_2(SO_4)_3 \cdot 9H_2O) = 562.0168$ g mol^{-1}
$M(Fe_2(SO_4)_3) = 399.8788$ g mol^{-1}

$$w = \frac{399.8788 \cdot 59.8}{562.0168}$$

$w = 42.55$ wt% $= w_1$

Preparation of 1 L of a solution with a concentration of 1 wt%:

$V_1 = ?$ mL
$V_1 + V_2 = 1$ L $= 1000$ mL
$\rho_1 = 1.538$ g cm^{-3}
$\rho_2 = 1$ g cm^{-3}
$w_1 = 42.55$ wt%
$w_2 = 0$
$w_3 = 1$ wt%
(Indexes: 1 = concentrated solution; 2 = water; 3 = diluted (1 wt%) solution)

$$m_1 w_1 + m_2 w_2 = (m_1 + m_2) w_3$$

$$V_1 \rho_1 w_1 + V_2 \rho_2 w_2 = (V_1 \rho_1 + V_2 \rho_2) w_3$$

$$V_1 \cdot 1.538 \cdot 42.55 + (1000 - V_1) \cdot 1 \cdot 0 = (V_1 \cdot 1.538 + (1000 - V_1) \cdot 1) \cdot 1$$

$$V_1 \cdot 1.538 \cdot 42.55 = V_1 \cdot 1.538 + 1000 - V_1$$

$$V_1 = 15.4074$$

Dilute **15.41 mL** of the concentrated stock solution to **1 L** with deionised water to prepare a **1 wt%** solution of $Fe_2(SO_4)_3$.

It is recommended not to shake the concentrated stock solution, and the desired stock solution volume should be pipetted from approximately 2 cm below the liquid surface. Preparation of a freshly diluted 1 wt% solution prior to conducting jar tests is suggested.

Calculating the volume of 1 wt% solution dosed into jars to achieve a desired Fe$_2$(SO$_4$)$_3$ dose:

The mass fraction of the solution $w = 0.01$ corresponds to 1 g of $Fe_2(SO_4)_3$ per 100 g of the solution. If the density of the 1 wt% solution is approximated by the density of water $\rho = 1$ g cm^{-3}, then 1 g of $Fe_2(SO_4)_3$ per 100 g of the solution corresponds to 10,000 mg of $Fe_2(SO_4)_3$ per 1 L of the solution. Thus, to reach, for example, a concentration of 10 mg L^{-1} $Fe_2(SO_4)_3$ in a jar, dilute the 1 wt% solution 1000 times.

Add **1 mL** of the 1 wt% solution of $Fe_2(SO_4)_3$ into **1 L** of raw water to achieve a dose of **10 mg L^{-1} Fe$_2$(SO$_4$)$_3$** in a jar.

Naturally, the required volume of 1 wt% solution may be easily recalculated for any other desired $Fe_2(SO_4)_3$ dose. It is recommended to always prepare a control sample (by dosing the calculated volume of 1 wt% $Fe_2(SO_4)_3$ solution into the corresponding volume of deionised water) in which the Fe concentration will be measured.

APPENDIX 5: PREPARATION OF 1% $AL_2(SO_4)_3$ SOLUTION

Stock solution:

$Al_2(SO_4)_3 \cdot 18H_2O$
$w = 49.8$ wt%
$\rho = 1.308$ g cm$^{-3} = \rho_1$
Conversion to anhydrous $Al_2(SO_4)_3$:
$M(Al_2(SO_4)_3 \cdot 18H_2O) = 666.4274$ g mol^{-1}
$M(Al_2(SO_4)_3) = 342.1538$ g mol^{-1}

$$w = \frac{342.1538 \cdot 49.8}{666.4274}$$

$w = 25.57$ wt% $= w_1$

Preparation of 1 L of a solution with a concentration of 1 wt%:

$V_1 = ?$ mL
$V_1 + V_2 = 1$ L $= 1000$ mL
$\rho_1 = 1.308$ g cm^{-3}
$\rho_2 = 1$ g cm^{-3}
$w_1 = 25.57$ wt%
$w_2 = 0$
$w_3 = 1$ wt%
(Indexes: 1 = concentrated solution; 2 = water; 3 = diluted (1 wt%) solution)

$$m_1 w_1 + m_2 w_2 = (m_1 + m_2) w_3$$

$$V_1 \rho_1 w_1 + V_2 \rho_2 w_2 = (V_1 \rho_1 + V_2 \rho_2) w_3$$

$$V_1 \cdot 1.308 \cdot 25.57 + (1000 - V_1) \cdot 1 \cdot 0 = (V_1 \cdot 1.308 + (1000 - V_1) \cdot 1) \cdot 1$$

$$V_1 \cdot 1.308 \cdot 25.57 = V_1 \cdot 1.308 + 1000 - V_1$$

$$V_1 = 30.1772$$

Dilute **30.18 mL** of the concentrated stock solution to **1 L** with deionised water to prepare a **1 wt%** solution of $Al_2(SO_4)_3$.

It is recommended not to shake the concentrated stock solution, and the desired stock solution volume should be pipetted from approximately 2 cm below the liquid surface. Preparation of a freshly diluted 1 wt% solution prior to conducting jar tests is suggested.

Calculating the volume of 1 wt% solution dosed into jars to achieve a desired $Al_2(SO_4)_3$ dose:

The mass fraction of the solution $w = 0.01$ corresponds to 1 g of $Al_2(SO_4)_3$ per 100 g of the solution. If the density of the 1 wt% solution is approximated by the density of water $\rho = 1$ g cm^{-3}, then 1 g of $Al_2(SO_4)_3$ per 100 g of the solution corresponds to 10,000 mg of $Al_2(SO_4)_3$ per 1 L of the solution. Thus, to reach, for example, a concentration of 10 mg L^{-1} $Al_2(SO_4)_3$ in a jar, dilute the 1 wt% solution 1000 times.

Add **1 mL** of the 1 wt% solution of $Al_2(SO_4)_3$ into **1 L** of raw water to achieve the dose of **10 mg L^{-1} $Al_2(SO_4)_3$** in a jar.

Naturally, the required volume of 1 wt% solution may be easily recalculated for any other desired $Al_2(SO_4)_3$ dose. It is recommended to always prepare a control sample (by dosing the calculated volume of 1 wt% $Al_2(SO_4)_3$ solution into the corresponding volume of deionised water) in which the Al concentration will be measured.

IWA Publishing's authorised EU representative for General Product
Safety Regulations is Diane D'Arras, 15 rue Duret, 75116 Paris,
France, e-mail: safety@iwap.co.uk.

Printed and bound by CPI Group (UK) Ltd, Croydon, CR0 4YY
06/07/2026
02157564-0003